Tucholsky Wagner Zola Scott Schlegel
Turgenev Wallace Fonatne Sydow Freud
Twain Walther von der Vogelweide Fouqué Friedrich II. von Preußen
Weber Freiligrath Frey
Fechner Fichte Weiße Rose von Fallersleben Kant Ernst Frommel
Richthofen
Engels Fielding Hölderlin
Fehrs Faber Flaubert Eichendorff Tacitus Dumas
Maximilian I. von Habsburg Fock Eliasberg Zweig Ebner Eschenbach
Feuerbach Ewald Eliot Vergil
Goethe Elisabeth von Österreich London
Mendelssohn Balzac Shakespeare Dostojewski Ganghofer
Trackl Stevenson Lichtenberg Rathenau Doyle Gjellerup
Mommsen Thoma Tolstoi Lenz Hambruch Hanrieder Droste-Hülshoff
Dach Verne von Arnim Hägele Hauff Humboldt
Karrillon Reuter Rousseau Hagen Hauptmann Gautier
Garschin
Damaschke Defoe Hebbel Baudelaire
Descartes
Hegel Kussmaul Herder
Wolfram von Eschenbach Darwin Dickens Schopenhauer Rilke George
Bronner Melville Grimm Jerome
Campe Horváth Aristoteles Bebel Proust
Bismarck Vigny Barlach Voltaire Federer Herodot
Gengenbach Heine
Storm Casanova Tersteegen Grillparzer Georgy
Brentano Chamberlain Lessing Langbein Gilm Gryphius
Strachwitz Claudius Schiller Lafontaine Kralik Iffland Sokrates
Katharina II. von Rußland Bellamy Schilling
Gerstäcker Raabe Gibbon Tschechow
Löns Hesse Hoffmann Gogol Wilde Gleim Vulpius
Luther Heym Hofmannsthal Klee Hölty Morgenstern Goedicke
Roth Heyse Klopstock Puschkin Homer Kleist
Luxemburg La Roche Horaz Mörike Musil
Machiavelli Kierkegaard Kraft Kraus
Navarra Aurel Musset Lamprecht Kind Kirchhoff Hugo Moltke
Nestroy Marie de France
Nietzsche Nansen Laotse Ipsen Liebknecht
Marx Lassalle Gorki Klett Leibniz Ringelnatz
von Ossietzky May vom Stein Lawrence Irving
Petalozzi Platon Knigge
Sachs Pückler Michelangelo Kock Kafka
Poe Liebermann Korolenko
de Sade Praetorius Mistral Zetkin

The publishing house tredition has created the series **TREDITION CLASSICS**. It contains classical literature works from over two thousand years. Most of these titles have been out of print and off the bookstore shelves for decades.

The book series is intended to preserve the cultural legacy and to promote the timeless works of classical literature. As a reader of a **TREDITION CLASSICS** book, the reader supports the mission to save many of the amazing works of world literature from oblivion.

The symbol of **TREDITION CLASSICS** is Johannes Gutenberg (1400 – 1468), the inventor of movable type printing.

With the series, tredition intends to make thousands of international literature classics available in printed format again – worldwide.

All books are available at book retailers worldwide in paperback and in hardcover. For more information please visit: www.tredition.com

tredition was established in 2006 by Sandra Latusseck and Soenke Schulz. Based in Hamburg, Germany, tredition offers publishing solutions to authors and publishing houses, combined with worldwide distribution of printed and digital book content. tredition is uniquely positioned to enable authors and publishing houses to create books on their own terms and without conventional manufacturing risks.

For more information please visit: www.tredition.com

No Animal Food and Nutrition and Diet with Vegetable Recipes

Rupert H. Wheldon

Imprint

This book is part of the TREDITION CLASSICS series.

Author: Rupert H. Wheldon
Cover design: toepferschumann, Berlin (Germany)

Publisher: tredition GmbH, Hamburg (Germany)
ISBN: 978-3-8491-5009-9

www.tredition.com
www.tredition.de

Copyright:
The content of this book is sourced from the public domain.

The intention of the TREDITION CLASSICS series is to make world literature in the public domain available in printed format. Literary enthusiasts and organizations worldwide have scanned and digitally edited the original texts. tredition has subsequently formatted and redesigned the content into a modern reading layout. Therefore, we cannot guarantee the exact reproduction of the original format of a particular historic edition. Please also note that no modifications have been made to the spelling, therefore it may differ from the orthography used today.

PREFACE

The title of this book is not ambiguous, but as it relates to a subject rarely thought about by the generality of people, it may save some misapprehension if at once it is plainly stated that the following pages are in vindication of a dietary consisting wholly of products of the vegetable kingdom, and which therefore excludes not only flesh, fish, and fowl, but milk and eggs and products manufactured therefrom.

The Author.

This work is reprinted from the English edition with changes better adapting it to the American reader.

The Publishers.

MAN'S FOOD

Health and happiness are within reach of those who provide themselves with good food, clean water, fresh air, and exercise.

A ceaseless and relentless hand is laid on almost every animal to provide food for human beings.

Nothing that lives or grows is missed by man in his search for food to satisfy his appetite.

Natural appetite is satisfied with vegetable food, the basis for highest and best health and development.

History of primitive man we know, but the possibilities of perfected and complete man are not yet attained.

Adequate and pleasant food comes to us from the soil direct, favorable for health, and a preventive against disease.

Plant food is man's natural diet; ample, suitable, and available; obtainable with least labor and expense, and in pleasing form and variety.

Animal food will be useful in emergency, also at other times; still, plant substance is more favorable to health, endurance, and power of mind.

Variety of food is desirable and natural; it is abundantly supplied by the growth of the soil under cultivation. [Pg 6]

Races of intelligence and strength are to be found subsisting and thriving on an exclusive plant grown diet.

The health and patience of vegetarians meet the social, mental and physical tests of life with less disease, and less risk of dependence in old age.

Meat eaters have no advantages which do not belong also to those whose food is vegetable.

Plant food, the principal diet of the world, has one serious drawback; it is not always savory, or palatable.

Plant diet to be savory requires fat, or oil, to be added to it; nuts, peanut, and olive oil, supply it to the best advantage.

Plant diet with butter, cream, milk, cheese, eggs, lard, fat, suet, or tallow added to it, is not vegetarian; it is mixed diet; the same in effect as if meat were used. — Elmer Lee, M.D., Editor, Health Culture Magazine. [Pg 7]

CONTENTS

No Animal Food

I — THE URGENCY OF THE SUBJECT
II — PHYSICAL CONSIDERATIONS
III — ETHICAL CONSIDERATIONS
IV — THE ÆSTHETIC POINT OF VIEW
V — ECONOMICAL CONSIDERATIONS
VI — THE EXCLUSION OF DAIRY PRODUCE
VII — CONCLUSION

Nutrition and Diet

I — SCIENCE OF NUTRITION
II — WHAT TO EAT
III — WHEN TO EAT
IV — HOW TO EAT

Food Table
Recipes

[Pg 8]
[Pg 9]

NO ANIMAL FOOD

I

URGENCY OF THE SUBJECT

Outside of those who have had the good fortune to be educated to an understanding of a rational science of dietetics, very few people indeed have any notion whatever of the fundamental principles of nutrition and diet, and are therefore unable to form any sound opinion as to the merits or demerits of any particular system of dietetic reform. Unfortunately many of those who *do* realise the intimate connection between diet and both physical and mental health, are not, generally speaking, sufficiently philosophical to base their views upon a secure foundation and logically reason out the whole problem for themselves.

Briefly, the pleas usually advanced on behalf of the vegetable regimen are as follows: It is claimed to be healthier than the customary flesh diet; it is claimed for various reasons to be more pleasant; it is claimed to be more economical; it is claimed to be less trouble; it is claimed to [Pg 10] be more humane. Many hold the opinion that a frugivorous diet is more natural and better suited to the constitution of man, and that he was never intended to be carnivorous; that the slaughtering of animals for food, being entirely unnecessary is immoral; that in adding our share towards supplying a vocation for the butcher we are helping to nurture callousness, coarseness and brutality in those who are concerned in the butchering business; that anyone of true refinement and delicacy would find in the killing of highly-strung, nervous, sensitive creatures, a task repulsive and disgusting, and that it is scarcely fair, let alone Christian, to ask others to perform work which we consider unnecessary and loathsome, and which we should be ashamed to do ourselves.

Of all these various views there is one that should be regarded as of primary importance, namely, the question of health. First and foremost we have to consider the question of physical health. No system of thought that poses as being concerned with man's welfare on earth can ever make headway unless it recognises this. Physical well-being is a moral consideration that should and must have our attention before aught else, and that this is so needs no demonstrating; it is self-evident.

Now it is not to be denied when we look at the over flowing hospitals; when we see everywhere advertised patent medicines; when we [Pg 11] realise that a vast amount of work is done by the medical profession among all classes; when we learn that one man out of twelve and one woman out of eight die every year from that most terrible disease, cancer, and that over 207,000 persons died from tuberculosis during the first seven years of the present century; when we learn that there are over 1500 defined diseases prevalent among us and that the list is being continually added to, that the general health of the nation is far different from what we have every reason to believe it ought to be. However much we may have become accustomed to it, we cannot suppose ill-health to be a *normal* condition. Granted, then, that the general health of the nation is far from what it should be, and looking from effects to causes, may we not pertinently enquire whether our diet is not largely responsible for this state of things? May it not be that wrong feeding and malnutrition are at the root of most disease? It needs no demonstrating that man's health is directly dependent upon what he eats, yet how few possess even the most elementary conception of the principles of nutrition in relation to health? Is it not evident that it is because of this lamentable ignorance so many people nowadays suffer from ill-health?

Further, not only does diet exert a definite influence upon physical well-being, but it indirectly affects the entire intellectual and moral evolution [Pg 12] of mankind. Just as a man thinks so he becomes, and 'a science which controls the building of brain-cell, and therefore of mind-stuff, lies at the root of all the problems of life.' From the point of view of food-science, mind and body are inseparable; one reacts upon the other; and though a healthy body may not be essential to happiness, good health goes a long way towards making life worth living. Dr. Alexander Haig, who has done such excellent and valuable work in the study of uric acid in relation to disease, speaks most emphatically on this point: 'DIET is the greatest question for the human race, not only does his ability to obtain food determine man's existence, but its quality controls the circulation in the brain, and this decides the trend of being and action, accounting for much of the indifference between depravity and the self-control of wisdom.'

The human body is a machine, not an iron and steel machine, but a blood and bone machine, and just as it is necessary to understand the mechanism of the iron and steel machine in order to run it, so is it necessary to understand the mechanism of the blood and bone machine in order to run it. If a person understanding nothing of the business of a *chauffeur* undertook to run an automobile, doubtless he would soon come to grief; and so likewise if a person understands nothing of the needs of his body, or partly understanding them knows not how to satisfy them, it [Pg 13] is extremely unlikely that he will maintain it at its normal standard of efficiency. Under certain conditions, of which we will speak in a moment, the body-machine is run quite unconsciously, and run well; that is to say, the body is kept in perfect health without the aid of science. But, then, we do not now live under these conditions, and so our reason has to play a certain part in encouraging, or, as the case may be, in restricting the various desires that make themselves felt. The reason so many people nowadays are suffering from all sorts of ailments is simply that they are deplorably ignorant of their natural bodily wants. How much does the ordinary individual know about nutrition, or about obedience to an unperverted appetite? The doctors seem to know little about health; they are not asked to keep us healthy, but only to cure us of disease, and so their studies relate to disease, not health; and dietetics, a science dealing with the very first principles of health, is an optional course in the curriculum of the medical student.

Food is the first necessary of life, and the right kind of food, eaten in the right manner, is necessary to a right, that is, healthy life. No doubt, pathological conditions are sometimes due to causes other than wrong feeding, but in a very large percentage of cases there is little doubt that errors in diet have been the cause of the trouble, either directly, or indirectly by rendering the system susceptible to pernicious [Pg 14] influences. [1] A knowledge of what is the right food to eat, and of the right way to eat it, does not, under existing conditions of life, come instinctively. Under other conditions it might do so, but under those in which we live, it certainly does not; and this is owing to the fact that for many hundred generations back there has been a pandering to sense, and a quelling and consequent atrophy of the discriminating animal instinct. As our intelligence has developed we have applied it to the service of the senses and at the expense of our primitive intuition of right and wrong that guided us in the selection of that which was suitable to our preservation and health. We excel the animals in the possession of reason, but the animals excel us in the exercise of instinct.

It has been said that animals do not study dietetics and yet live healthily enough. This is true, but it is true only as far as concerns those [Pg 15] animals which live *in their natural surroundings and under natural conditions*. Man would not need to study diet were he so situated, but he is not. The wild animal of the woods is far removed from the civilized human being. The animal's instinct guides him aright, but man has lost his primitive instinct, and to trust to his inclinations may result in disaster.

The first question about vegetarianism, then, is this:—Is it the best diet from the hygienic point of view? Of course it will be granted that diseased food, food containing pernicious germs or poisons, whether animal or vegetable, is unfit to be eaten. It is not to be supposed that anyone will defend the eating of such food, so that we are justified in assuming that those who defend flesh-eating believe flesh to be free from such germs and poisons; therefore let the following be noted. It is affirmed that 50 per cent. of the bovine and other animals that are slaughtered for human food are affected with Tuberculosis, or some of the following diseases: Cancer, Anthrax, Pleuro-Pneumonia, Swine-Fever, Sheep Scab, Foot and Mouth Disease, etc., etc., and that to exclude all suspected or actually diseased carcasses would be practically to leave the market without a supply. One has only to read the literature dealing with this subject to be convinced that the meat-eating public must consume a large amount of highly poisonous substances. That these poisons may communicate disease to [Pg 16] the person eating them has been amply proved. Cooking does *not* necessarily destroy all germs, for the temperature at the interior of a large joint is below that necessary to destroy the bacilli there present.

Although the remark is irrelevant to the subject in hand, one is tempted to point out that, quite apart from the question of hygiene, the idea of eating flesh containing sores and wounds, bruises and pus-polluted tissues, is altogether repulsive to the imagination.

Let it be supposed, however, that meat can be, and from the meat-eater's point of view, should be and will be under proper conditions, uncontaminated, there yet remains the question whether such food is physiologically necessary to man. Let us first consider what kind of food is best suited to man's natural constitution. [Pg 17]

II

PHYSICAL CONSIDERATIONS

There are many eminent scientists who have given it as their opinion that anatomically and physiologically man is to be classed as a frugivorous animal. There are lacking in man all the characteristics that distinguish the prominent organs of the carnivora, while he possesses a most striking resemblance to the fruit-eating apes. Dr. Kingsford writes: 'M. Pouchet observes that all the details of the digestive apparatus in man, as well as his dentition, constitute "so many proofs of his frugivorous origin" — an opinion shared by Professor Owen, who remarks that the anthropoids and all the quadrumana derive their alimentation from fruits, grains, and other succulent and nutritive vegetable substances, and that the strict analogy which exists between the structure of these animals and that of man clearly demonstrates his frugivorous nature. This view is also taken by Cuvier, Linnæus, Professor Lawrence, Charles Bell, Gassendi, Flourens, and a great number of other eminent writers.' (see *The Perfect Way in Diet*.)

Linnæus is quoted by John Smith in *Fruits* [Pg 18] *and Farinacea* as speaking of fruit as follows: 'This species of food is that which is most suitable to man: which is evidenced by the series of quadrupeds, analogy, wild men, apes, the structure of the mouth, of the stomach, and the hands.'

Sir Ray Lancaster, K.C.B., F.R.S., in an article in *The Daily Telegraph*, December, 1909, wrote: 'It is very generally asserted by those who advocate a purely vegetable diet that man's teeth are of the shape and pattern which we find in the fruit-eating, or in the root-eating, animals allied to him. This is true.... It is quite clear that man's cheek teeth do not enable him to cut lumps of meat and bone from raw carcasses and swallow them whole. They are broad, square-surfaced teeth with four or fewer low rounded tubercles to crush soft food, as are those of monkeys. And there can be no doubt that man fed originally like monkeys, on easily crushed fruits, nuts, and roots.'

With regard to man's original non-carnivorous nature and omnivorism, it is sometimes said that though man's system may not thrive on a raw flesh diet, yet he can assimilate cooked flesh and his system is well adapted to digest it. The answer to this is that were it demonstrable, and it is *not*, that cooked flesh is as easily digested and contains as much nutriment as grains and nuts, this does not prove it to be suitable for human food; for man (leaving out [Pg 19] of consideration the fact that the eating of diseased animal flesh can communicate disease), since he was originally formed by Nature to subsist exclusively on the products of the vegetable kingdom, cannot depart from Nature's plan without incurring penalty of some sort—unless, indeed, his natural original constitution has changed; but *it has not changed*. The most learned and world-renowned scientists affirm man's present anatomical and physiological structure to be that of a frugivore. Disguising an unnatural food by cooking it may make that food more assimilable, but it by no means follows that such a food is suitable, let alone harmless, as human food. That it is harmful, not only to man's physical health, but to his mental and moral health, this book endeavours to demonstrate.

With regard to the fact that man has not changed constitutionally from his original frugivorous nature Dr. Haig writes as follows: 'If man imagines that a few centuries, or even a few hundred centuries, of meat-eating in defiance of Nature have endowed him with any new powers, except perhaps, that of bearing the resulting disease and degradation with an ignorance and apathy which are appalling, he deceives himself; for the record of the teeth shows that human structure has remained unaltered over vast periods of time.'

According to Dr. Haig, human metabolism (the process by which food is converted into [Pg 20] living tissue) differs widely from that of the carnivora. The carnivore is provided with the means to dispose of such poisonous salts as are contained in and are produced by the ingestion of animal flesh, while the human system is not so provided. In the human body these poisons are not held in solution, but tend to form deposits and consequently are the cause of diseases of the arthritic group, conspicuously rheumatism.

There is sometimes some misconception as regards the distinction between a frugivorous and herbivorous diet. The natural diet of man consists of fruits, farinacea, perhaps certain roots, and the more esculent vegetables, and is commonly known as vegetarian, or fruitarian (frugivorous), but man's digestive organs by no means allow him to eat grass as the herbivora—the horse, ox, sheep, etc.—although he is much more nearly allied to these animals than to the carnivora.

We are forced to conclude, in the face of all the available evidence, that the natural constitution of man closely resembles that of fruit-eating animals, and widely differs from that of flesh-eating animals, and that from analogy it is only reasonable to suppose that the fruitarian, or vegetarian, as it is commonly called, is the diet best suited to man. This conclusion has been arrived at by many distinguished men of science, among whom are the above mentioned. But the proof of the pudding is in the eating, [Pg 21] and to prove that the vegetarian is the most hygienic diet, we must examine the physical conditions of those nations and individuals who have lived, and do live, upon this diet.

It might be mentioned, parenthetically, that among animals, the herbivora are as strong physically as any species of carnivora. The most laborious work of the world is performed by oxen, horses, mules, camels, elephants, all vegetable-feeding animals. What animal possesses the enormous strength of the herbivorous rhinoceros, who, travellers relate, uproots trees and grinds whole trunks to powder? Again, the frugivorous orang-outang is said to be more than a match for the African lion. Comparing herbivora and carnivora from this point of view Dr. Kingsford writes: 'The carnivora, indeed, possess one salient and terrible quality, ferocity, allied to thirst for blood; but power, endurance, courage, and intelligent capacity for toil belong to those animals who alone, since the world has had a history, have been associated with the fortunes, the conquests, and the achievements of men.'

Charles Darwin, reverenced by all educated people as a scientist of the most keen and accurate observation, wrote in his *Voyage of the Beagle*, the following with regard to the Chilian miners, who, he tells us, live in the cold and high regions of the Andes: 'The labouring class work very hard. They have little time allowed for their meals, and during summer and winter, [Pg 22] they begin when it is light and leave off at dusk. They are paid £1 sterling a month and their food is given them: this, for breakfast, consists of sixteen figs and two small loaves of bread; for dinner, boiled beans; for supper, broken roasted wheat-grain. They scarcely ever taste meat.' This is as good as saying that the strongest men in the world, performing the most arduous work, and living in an exhilarating climate, are practically strict vegetarians.

Dr. Jules Grand, President of the Vegetarian Society of France speaks of 'the Indian runners of Mexico, who offer instances of wonderful endurance, and eat nothing but tortillas of maize, which they eat as they run along; the street porters of Algiers, Smyrna, Constantinople and Egypt, well known for their uncommon strength, and living on nothing but maize, rice, dates, melons, beans, and lentils. The Piedmontese workmen, thanks to whom the tunnelling of the Alps is due, feed on polenta, (maize-broth). The peasants of the Asturias, like those of the Auvergne, scarcely eat anything except chick-peas and chestnuts ... statistics prove ... that the most numerous population of the globe is vegetarian.'

The following miscellaneous excerpta are from Smith's *Fruits and Farinacea*:—

'The peasantry of Norway, Sweden, Russia, Denmark, Poland, Germany, Turkey, Greece, Switzerland, Spain, Portugal, and of almost every country in Europe subsist principally, and most [Pg 23] of them entirely, on vegetable food.... The Persians, Hindoos,

Burmese, Chinese, Japanese, the inhabitants of the East Indian Archipelago, and of the mountains of the Himalaya, and, in fact, most of the Asiatics, live upon vegetable productions.'

'The people of Russia, generally, subsist on coarse black rye-bread and garlics. I have often hired men to labour for me. They would come on board in the morning with a piece of black bread weighing about a pound, and a bunch of garlics as big as one's fist. This was all their nourishment for the day of sixteen or eighteen hours' labour. They were astonishingly powerful and active, and endured severe and protracted labour far beyond any of my men. Some of these Russians were eighty and even ninety years old, and yet these old men would do more work than any of the middle-aged men belonging to my ship. Captain C. S. Howland of New Bedford, Mass.'

'The Chinese feed almost entirely on rice, confections and fruits; those who are enabled to live well and spend a temperate life, are possessed of great strength and agility.'

'The Egyptian cultivators of the soil, who live on coarse wheaten bread, Indian corn, lentils, and other productions of the vegetable kingdom, are among the finest people I have even seen. Latherwood.'

'The Greek boatmen are exceedingly abstem [Pg 24] ious. Their food consists of a small quantity of black bread, made of unbolted rye or wheatmeal, and a bunch of grapes, or raisins, or some figs. They are astonishingly athletic and powerful; and the most nimble, active, graceful, cheerful, and even merry people in the world. Judge Woodruff, of Connecticut.'

'From the day of his irruption into Europe the Turk has always proved himself to be endowed with singularly strong vitality and energy. As a member of a warlike race, he is without equal in Europe in health and hardiness. His excellent physique, his simple habits, his abstinence from intoxicating liquors, and his normal vegetarian diet, enable him to support the greatest hardships, and to exist on the scantiest and simplest food.'

'The Spaniards of Rio Salada in South America,—who come down from the interior, and are employed in transporting goods overland,—live wholly on vegetable food. They are large, very robust, and strong; and bear prodigious burdens on their backs, travelling over mountains too steep for loaded mules to ascend, and with a speed which few of the generality of men can equal without incumbrance.'

'In the most heroic days of the Grecian army, their food was the plain and simple produce of the soil. The immortal Spartans of Thermopylæ were, from infancy, nourished by the plainest and coarsest vegetable aliment: and the Roman [Pg 25] army, in the period of their greatest valour and most gigantic achievements, subsisted on plain and coarse vegetable food. When the public games of Ancient Greece—for the exercise of muscular power and activity in wrestling, boxing, running, etc.,—were first instituted, the athletæ in accordance with the common dietetic habits of the people, were trained entirely on vegetable food.'

Dr. Kellogg, an authority on dietetics, makes the following answer to those who proclaim that those nations who eat a large amount of flesh-food, such as the English, are the strongest and dominant nations: "While it is true that the English nation makes large use of animal food, and is at the same time one of the most powerful on the globe, it is also true that the lowest, most miserable classes of human beings, such as the natives of Australia, and the inhabitants of Terra del Fuego, subsist almost wholly upon flesh. It should also be borne in mind that it is only within a single generation that the common people of England have become large consumers of flesh. In former times and when England was laying the foundation of her greatness, her sturdy yeomen ate less meat in a week, than the average Englishman of the present consumes in a single day.... The Persians, the Grecians, and the Romans, became ruling nations while vegetarians."

In *Fruits and Farinacea*, Professor Lawrence [Pg 26] is quoted as follows: 'The inhabitants of Northern Europe and Asia, the Laplanders, Samoiedes, Ostiacs, Tangooses, Burats, Kamtschatdales, as well as the natives of Terra del Fuego in the Southern extremity of America, are the smallest, weakest, and least brave people on the globe; although they live almost entirely on flesh, and that often raw.'

Many athletic achievements of recent date have been won by vegetarians both in this country and abroad. The following successes are noteworthy: — Walking: Karl Mann, Dresden to Berlin, Championship of Germany; George Allen, Land's End to John-o'-Groats. Running: E. R. Voigt, Olympic Championship, etc.: F. A. Knott, 5,000 metres Belgian record. Cycling: G. A. Olley, Land's End to John-o'-Groats record. Tennis: Eustace Miles, M.A., various championships, etc. Of especial interest at the present moment are a series of tests and experiments recently carried out at Yale University, U.S.A., under Professor Irving Fisher, with the object of discovering the suitability of different dietaries for athletes, and the effect upon the human system in general. The results were surprising. 'One of the most severe tests,' remarks Professor Fisher, 'was in deep knee-bending, or "squatting." Few of the meat-eaters could "squat" more than three to four hundred times. On the other hand a Yale student who had been a flesh-abstainer for two years, did the deep knee-bending eighteen hundred times [Pg 27] without exhaustion.... One remarkable difference between the two sets of men was the comparative absence of soreness in the muscles of the meat-abstainers after the tests.'

The question as to climate is often raised; many people labour under the idea that a vegetable diet may be suitable in a hot climate, but not in a cold. That this idea is false is shown by facts, some of which the above quotations supply. That man can live healthily in arctic regions on a vegetable diet has been amply demonstrated. In a cold climate the body requires a considerable quantity of heat-producing food, that is, food containing a good supply of hydrocarbons (fats), and carbohydrates (starches and sugars). Many vegetable foods are rich in these properties, as will be explained in the essay following dealing with dietetics. Strong and enduring vegetable-feeding animals, such as the musk-ox and the reindeer, flourish on the scantiest food in an arctic climate, and there is no evidence to show that man could not equally well subsist on vegetable food under similar conditions.

In an article entitled *Vegetarianism in Cold Climates*, by Captain Walter Carey, R.N., the author describes his observations during a winter spent in Manchuria. The weather, we are told, was exceedingly cold, the thermometer falling as low as minus 22° F. After speaking of the various arduous labours the natives are engaged in, Captain Carey describes the physique and diet [Pg 28] of natives in the vicinity of Niu-Chwang as follows: 'The men accompanying the carts were all very big and of great strength, and it was obvious that none but exceptionally strong and hardy men could withstand the hardships of their long march, the intense cold, frequent blizzards, and the work of forcing their queer team along in spite of everything. One could not help wondering what these men lived on, and I found that the chief article was beans, which, made into a coarse cake, supplied food for both men and animals. I was told by English merchants who travelled in the interior, that everywhere they found the same powerful race of men, living on beans and rice—in fact, vegetarians. Apparently they obtain the needful proteid and fat from the beans; while the coarse once-milled rice furnishes them with starch, gluten, and mineral salts, etc. Spartan fare, indeed, but proving how easy it is to sustain life without consuming flesh-food.'

So far, then, as the physical condition of those nations who are practically vegetarian is concerned, we have to conclude that practice tallies with theory. Science teaches that man should live on a non-flesh diet, and when we come to consider the physique of those nations and men who do so, we have to acknowledge that their bodily powers and their health equal, if not excel, those of nations and men who, in part, subsist upon flesh. But it is interesting to go yet further. [Pg 29] It has already been stated that mind and body are inseparable; that one reacts upon the other: therefore it is not irrelevant, in passing, to observe what mental powers are possessed by those races and individuals who subsist entirely upon the products of the vegetable kingdom.

When we come to consider the mentality of the Oriental races we certainly have to acknowledge that Oriental culture—ethical, metaphysical, and poetical—has given birth to some of the grandest and noblest thoughts that mankind possesses, and has devised philosophical systems that have been the comfort and salvation of countless millions of souls. Anyone who doubts the intellectual and ethical attainments of that remarkable nation of which we in the West know so little—the Chinese—should read the panegyric written by Sir Robert Hart, who, for forty years, lived among them, and learnt to love and venerate them as worthy of the highest admiration and respect. Others have written in praise of the people of Burma. Speaking of the Burman, a traveller writes: 'He will exercise a graceful charity unheard of in the West—he has discovered how to make life happy without selfishness and to combine an adequate power for hard work with a corresponding ability to enjoy himself gracefully ... he is a philosopher and an artist.'

Speaking of the Indian peasant a writer in an English journal says: 'The ryot lives in the face of Nature, on a simple diet easily procured, and [Pg 30] inherits a philosophy, which, without literary culture, lifts his spirit into a higher plane of thought than other peasantries know of. Abstinence from flesh food of any kind, not only gives him pure blood exempt from civilized diseases but makes him the friend and not the enemy, of the animal world around.'

Eastern literature is renowned for its subtle metaphysics. The higher types of Orientals are endowed with an extremely subtle intelligence, so subtle as to be wholly unintelligible to the ordinary Westerner. It is said that Pythagoras and Plato travelled in the East and were initiated into Eastern mysticism. The East possesses many scriptures, and the greater part of the writings of Eastern scholars consist of commentaries on the sacred writings. Among the best known monumental philosophical and literary achievements maybe mentioned the *Tao Teh C'hing*; the *Zend Avesta*; the *Three Vedas*; the *Brahmanas*; the *Upanishads*; and the *Bhagavad-gita*, that most beautiful 'Song Celestial' which for nearly two thousand years has moulded the thoughts and inspired the aspirations of the teeming millions of India.

As to the testimony of individuals it is interesting to note that some of the greatest philosophers, scientists, poets, moralists, and many men of note, in different walks of life, in past and modern times, have, for various reasons, been [Pg 31] vegetarians, among whom have been named the following:—

- Manu
- Zoroaster
- Pythagoras
- Zeno

- Buddha
- Isaiah
- Daniel
- Empedocles
- Socrates
- Plato
- Aristotle
- Porphyry
- John Wesley
- Franklin
- Goldsmith
- Ray
- Paley
- Isaac Newton
- Jean Paul Richter
- Schopenhauer
- Byron
- Gleizes
- Hartley
- Rousseau
- Iamblichus
- Hypatia
- Diogenes
- Quintus Sextus
- Ovid
- Plutarch
- Seneca
- Apollonius
- The Apostles
- Matthew
- James
- James the Less
- Peter
- The Christian Fathers
- Clement
- Tertullian

- Origen
- Chrysostom
- St. Francis d'Assisi
- Cornaro
- Leonardo da Vinci
- Milton
- Locke
- Spinoza
- Voltaire
- Pope
- Gassendi
- Swedenborg
- Thackeray
- Linnæus
- Shelley
- Lamartine
- Michelet
- William Lambe
- Sir Isaac Pitman
- Thoreau
- Fitzgerald
- Herbert Burrows
- Garibaldi
- Wagner
- Edison
- Tesla
- Marconi
- Tolstoy
- George Frederick Watts
- Maeterlinck
- Vivekananda
- General Booth
- Mrs. Besant
- Bernard Shaw
- Rev. Prof. John E. B. Mayor
- Hon. E. Lyttelton

- Rev. R. J. Campbell
- Lord Charles Beresford
- Gen. Sir Ed. Bulwer
- etc., etc., etc.

[Pg 32]

The following is a list of the medical and scientific authorities who have expressed opinions favouring vegetarianism:—

- M. Pouchet
- Baron Cuvier
- Linnæus
- Professor Laurence, F.R.S.
- Sir Charles Bell, F.R.S.
- Gassendi
- Flourens
- Sir John Owen
- Professor Howard Moore
- Sylvester Graham, M.D.
- John Ray, F.R.S.
- Professor H. Schaafhausen
- Sir Richard Owen, F.R.S.
- Charles Darwin, LL.D., F.R.S.
- Dr. John Wood, M.D.
- Professor Irving Fisher
- Professor A. Wynter Blyth, F.R.C.S.
- Edward Smith, M.B., F.R.S., LL.B.
- Adam Smith, F.R.S.
- Lord Playfair, M.D., C.B.
- Sir Henry Thompson, M.B., F.R.C.S.
- Dr. F. J. Sykes, B. Sc.
- Dr. Anna Kingsford
- Professor G. Sims Woodhead, M.D., F.R.C.P., F.R.S.
- Alexander Haig, M.A., M.D., F.R.C.P.
- Dr. W. B. Carpenter, C.B., F.R.S.

- Dr. Josiah Oldfield, D.C.L., M.A., M.R.C.S., L.R.C.P.
- Virchow
- Sir Benjamin W. Richardson, M.P., F.R.C.S.
- Dr. Robert Perks, M.D., F.R.C.S.
- Dr. Kellogg, M.D.
- Harry Campbell, M.D.
- Dr. Olsen
- etc., etc.

Before concluding this section it might be pointed out that the curious prejudice which is always manifested when men are asked to consider any new thing is as strongly in evidence against food reform as in other innovations. For example, flesh-eating is sometimes defended on the ground that vegetarians do not look hale and hearty, as healthy persons should do. People who speak in this way probably have in mind one or two acquaintances who, through [Pg 33] having wrecked their health by wrong living, have had to abstain from the 'deadly decoctions of flesh' and adopt a simpler and purer dietary. It is not fair to judge meat abstainers by those who have had to take to a reformed diet solely as a curative measure; nor is it fair to lay the blame of a vegetarian's sickness on his diet, as if it were impossible to be sick from any other cause. The writer has known many vegetarians in various parts of the world, and he fails to understand how anyone moving about among vegetarians, either in this country or elsewhere, can deny that such people look as healthy and cheerful as those who live upon the conventional omnivorous diet.

If a vegetarian, owing to inherited susceptibilities, or incorrect rearing in childhood, or any other cause outside his power to prevent, is sickly and delicate, is it just to lay the blame on his present manner of life? It would, indeed, seem most reasonable to assume that the individual in question would be in a much worse condition had he not forsaken his original and mistaken diet when he did. The writer once heard an acquaintance ridicule vegetarianism on the ground that Thoreau died of pulmonary consumption at forty-five! One is reminded of Oliver Wendell Holmes' witty saying:—'The mind of the bigot is like the pupil of the eye: the more it sees the light, the more it contracts.' [Pg 34]

In conclusion, there is, as we have seen in our review of typical vegetarian peoples and classes throughout the world, the strongest evidence that those who adopt a sensible non-flesh dietary, suited to their own constitution and environment, are almost invariably healthier, stronger, and longer-lived than those who rely chiefly upon flesh-meat for nutriment. [Pg 35]

III

ETHICAL CONSIDERATIONS

The primary consideration in regard to the question of diet should be, as already stated, the hygienic. Having shown that the non-flesh diet is the more natural, and the more advantageous from the point of view of health, let us now consider which of the two — vegetarianism or omnivorism — is superior from the ethical point of view.

The science of ethics is the science of conduct. It is founded, primarily, upon philosophical postulates without which no code or system of morals could be formulated. Briefly, these postulates are, (a), every activity of man has as its deepest motive the end termed Happiness, (b) the Happiness of the individual is indissolubly bound up with the Happiness of all Creation. The truth of (a) will be evident to every person of normal intelligence: all arts and systems aim consciously, or unconsciously, at some good, and so far as names are concerned everyone will be willing to call the Chief Good by the term Happiness, al [Pg 36] though there may be unlimited diversity of opinion as to its nature, and the means to attain it. The truth of (b) also becomes apparent if the matter is carefully reflected upon. Everything that is *en rapport* with all other things: the pebble cast from the hand alters the centre of gravity in the Universe. As in the world of things and acts, so in the world of thought, from which all action springs. Nothing can happen to the part but the whole gains or suffers as a consequence. Every breeze that blows, every cry that is uttered, every thought that is born, affects through perpetual metamorphoses every part of the entire Cosmic Existence. [2]

We deduce from these postulates the following ethical precepts: a wise man will, firstly, so regulate his conduct that thereby he may experience the greatest happiness; secondly, he will endeavour to bestow happiness on others that by so doing he may receive, indirectly, being himself a part of the Cosmic Whole, the happiness he gives. Thus supreme selfishness is synonymous with supreme egoism, a truth that can only be stated paradoxically.

Applying this latter precept to the matter in hand, it is obvious that since we should so live as to give the greatest possible happiness to all [Pg 37] beings capable of appreciating it, and as it is an indisputable fact that animals can suffer pain, *and that men who slaughter animals needlessly suffer from atrophy of all finer feelings*, we should therefore cause no unnecessary suffering in the animal world. Let us then consider whether, knowing flesh to be unnecessary as an article of diet, we are, in continuing to demand and eat flesh-food, acting morally or not. To answer this query is not difficult.

It is hardly necessary to say that we are causing a great deal of suffering among animals in breeding, raising, transporting, and killing them for food. It is sometimes said that animals do not suffer if they are handled humanely, and if they are slaughtered in abattoirs under proper superintendence. But we must not forget the branding and castrating operations; the journey to the slaughterhouse, which when trans-continental and trans-oceanic must be a long drawn-out nightmare of horror and terror to the doomed beasts; we must not forget the insatiable cruelty of the average cowboy; we must not forget that the animal inevitably spends at least some minutes of instinctive dread and fear when he smells and sees the spilt blood of his forerunners, and that this terror is intensified when, as is frequently the case, he witnesses the dying struggles, and hears the heart-rending groans; we must not forget that the best contrivances sometimes fail to do good work, and that a certain percentage [Pg 38] of victims have to suffer a prolonged death-agony owing to the miscalculation of a bad workman. Most people go through life without thinking of these things: they do not stop and consider from whence and by what means has come to their table the flesh-food that is served there. They drift along through a mundane existence without feeling a pang of remorse for, or even thought of, the pain they are accomplices in producing in the sub-human world. And it cannot be denied, hide it how we may, either from our eyes or our conscience, that however skilfully the actual killing may usually be carried out, there is much unavoidable suffering caused to the beasts that have to be transported by sea and rail to the slaughter-house. The animals suffer violently from sea-sickness, and horrible cruelty (such as pouring boiling oil into their ears, and stuffing their ears with hay which is then set on fire, tail-twisting, etc.,) has to be practised to prevent them lying down lest they be trampled on by other beasts and killed; for this means that they have to be thrown overboard, thus reducing the profits of their owners, or of the insurance companies, which, of course, would be a sad calamity. Judging by the way the men act it does not seem to matter what cruelties and tortures are perpetuated; what heinous offenses against every humane sentiment of the human heart are committed; it does not matter to what depths of Satanic callousness man stoops provided always that — this is [Pg 39] the supreme question — *there is money to be made by it.*

A writer has thus graphically described the scene in a cattle-boat in rough weather: 'Helpless cattle dashed from one side of the ship to the other, amid a ruin of smashed pens, with limbs broken from contact with hatchway combings or winches — dishorned, gored, and some of them smashed to mere bleeding masses of hide-covered flesh. Add to this the shrieking of the tempest, and the frenzied moanings of the wounded beasts, and the reader will have some faint idea of the fearful scenes of danger and carnage ... the dead beasts, advanced, perhaps, in decomposition before death ended their sufferings, are often removed literally in pieces.'

And on the railway journey, though perhaps the animals do not experience so much physical pain as travelling by sea, yet they are often deprived of food, and water, and rest, for long periods, and mercilessly knocked about and bruised. They are often so injured that the cattle-men are surprised they have not succumbed to their injuries. And all this happens in order that the demand for *unnecessary* flesh-food may be satisfied.

Those who defend flesh-eating often talk of humane methods of slaughtering; but it is significant that there is considerable difference of opinion as to what *is* the most humane method. In England the pole-axe is used; in Germany the [Pg 40] mallet; the Jews cut the throat; the Italians stab. It is obvious that each of these methods cannot be better than the others, yet the advocates of each method consider the others cruel. As Lieut. Powell remarks, this 'goes far to show that a great deal of cruelty and suffering is inseparable from all methods.'

It is hard to imagine how anyone believing he could live healthily on vegetable food alone, could, having once considered these things, continue a meat-eater. At least to do so he could not live his life in conformity with the precept that we should cause no unnecessary pain.

> How unholy a custom, how easy a way to murder he makes for himself
>
> Who cuts the innocent throat of the calf, and hears unmoved its mournful plaint!
>
> And slaughters the little kid, whose cry is like the cry of a child,
>
> Or devours the birds of the air which his own hands have fed!
>
> Ah, how little is wanting to fill the cup of his wickedness!
>
> What unrighteous deed is he not ready to commit.

> Make war on noxious creatures, and kill them only,
>
> But let your mouths be empty of blood, and satisfied with pure and natural repasts.

Ovid. *Metam.*, lib. xv.

That we cannot find any justification for destroying animal life for food does not imply we should never destroy animal life. Such a cult would be pure fanaticism. If we are to consider physical well-being as of primary importance, it follows that we shall act in self-preservation [Pg 41] 'making war on noxious creatures.' But this again is no justification for 'blood-sports.'

He who inflicts pain needlessly, whether by his own hand or by that of an accomplice, not only injures his victim, but injures himself. He stifles what nobleness of character he may have and he cultivates depravity and barbarism. He destroys in himself the spirit of true religion and isolates himself from those whose lives are made beautiful by sympathy. No one need hope for a spiritual Heaven while helping to make the earth a bloody Hell. No one who asks others to do wrong for him need imagine he escapes the punishment meted out to wrong-doers. That he procures the service of one whose sensibilities are less keen than his own to procure flesh-food for him that he may gratify his depraved taste and love of conformity does not make him less guilty of crime. Were he to kill with his own hand, and himself dress and prepare the obscene food, the evil would be less, for then he would not be an accomplice in retarding the spiritual growth of a fellow being. There is no shame in any *necessary* labour, but that which is unnecessary is unmoral, and slaughtering animals to eat their flesh is not only unnecessary and unmoral; it is also cruel and immoral. Philosophers and transcendentalists who believe in the Buddhist law of Kârma, Westernized by Emerson and Carlyle into the great doctrine of Compensation, realize that every act of unkindness, every deed that is con [Pg 42] trary to the dictates of our nobler instincts and reason, reacts upon us, and we shall truly reap that which we have sown. An act of brutality brutalizes, and the more we become brutalized the more we attract natures similarly brutal and get treated by them brutally. Thus does Nature sternly deal justice.

'Our acts our angels are, or good or ill,

Our fatal shadows that walk by us still.'

It is appropriate in this place to point out that some very pointed things are said in the Bible against the killing and eating of animals. It has been said that it is possible by judiciously selecting quotations to find the Bible support almost anything. However this may be, the following excerpta are of interest:—

'And God said: Behold, I have given you every herb bearing seed, and every tree in which is the fruit of a tree yielding seed, to you it shall be for meat.'—Gen. i., 29.

'But flesh with life thereof, which is the blood thereof, ye shall not eat.'—Gen. ix., 4.

'It shall be a perpetual statute throughout your generations in all your dwellings, that ye shall eat neither fat nor blood.'—Lev. iii., 17.

'Ye shall eat no manner of blood, whether it be of fowl, or beast.'—Lev. vii., 26.

'Ye shall eat the blood of no manner of flesh: for the life of all flesh is the blood thereof: whosoever eateth it shall be cut off.'—Lev. xvii., 14. [Pg 43]

'The wolf also shall dwell with the lamb, and the leopard shall lie down with the kid; and the calf and the young lion and the fatling together; and a little child shall lead them.... They shall not hurt nor destroy in all my holy mountain.'—Isaiah lxv.

'He that killeth an ox is as he that slayeth a man.'—Isaiah lxvi., 3.

'I desire mercy, and not sacrifice.'—Matt. ix., 7.

'It is good not to eat flesh, nor to drink wine, nor to do anything whereby thy brother stumbleth.'—Romans xiv., 21.

'Wherefore, if meat maketh my brother to stumble I will eat no flesh for evermore, that I make not my brother stumble.'—1 Cor. viii., 13.

The verse from Isaiah is no fanciful stretch of poetic imagination. The writer, no doubt, was picturing a condition of peace and happiness on earth, when discord had ceased and all creatures obeyed Nature and lived in harmony. It is not absurd to suppose that someday the birds and beasts may look upon man as a friend and benefactor, and not the ferocious beast of prey that he now is. In certain parts of the world, at the present day—the Galapagos Archipelago, for instance—where man has so seldom been that he is unknown to the indigenous animal life, travellers relate that birds are so tame and friendly and curious, being wholly unacquainted with the bloodthirsty nature of man, that they will perch [Pg 44] on his shoulders and peck at his shoe laces as he walks.

It may be said that Jesus did not specifically forbid flesh-food. But then he did not specifically forbid war, sweating, slavery, gambling, vivisection, cock and bull fighting, rabbit-coursing, trusts, opium smoking, and many other things commonly looked upon as evils which should not exist among Christians. Jesus laid down general principles, and we are to apply these general principles to particular circumstances.

The sum of all His teaching is that love is the most beautiful thing in the world; that the Kingdom of Heaven is open to all who really and truly love. The act of loving is the expression of a desire to make others happy. All beings capable of experiencing pain, who have nervous sensibilities similar to our own, are capable of experiencing the effect of our love. The love which is unlimited, which is not confined merely to wife and children, or blood relations and social companions, or one's own nation, or even the entire human race, but is so comprehensive as to include all life, human and sub-human; such love as this marks the highest point in moral evolution that human intelligence can conceive of or aspire to.

Eastern religions have been more explicit than Christianity about the sin of killing animals for food.

In the *Laws of Manu*, it is written: 'The man who forsakes not the law, and eats not flesh-meat [Pg 45] like a bloodthirsty demon, shall attain goodness in this world, and shall not be afflicted with maladies.'

'Unslaughter is the supreme virtue, supreme asceticism, golden truth, from which springs up the germ of religion.' *The Mahabharata.*

'*Non-killing*, truthfulness, non-stealing, continence, and non-receiving, are called Yama.' *Patanjalis' Yoga Aphorisms.*

'A Yogî must not think of injuring anyone, through thought, word or deed, and this applies not only to man, but to all animals. Mercy shall not be for men alone, but shall go beyond, and embrace the whole world.' *Commentary of Vivekânanda.*

'Surely hell, fire, and repentance are in store for those who for their pleasure and gratification cause the dumb animals to suffer pain.' *The Zend Avesta.*

Gautama, the Buddha, was most emphatic in discountenancing the killing of animals for food, or for any other unnecessary purpose, and Zoroaster and Confucius are said to have taught the same doctrine. [Pg 46]

IV

THE ÆSTHETIC POINT OF VIEW

St. Paul tells us to think on whatsoever things are pure and lovely (Phil. iv., 8). The implication is that we should love and worship beauty. We should seek to surround ourselves by beautiful objects and avoid that which is degrading and ugly.

Let us make some comparisons. Look at a collection of luscious fruits filling the air with perfume, and pleasing the eye with a harmony of colour, and then look at the gruesome array of skinned carcasses displayed in a butcher's shop; which is the more beautiful? Look at the work of the husbandman, tilling the soil, pruning the trees, gathering in the rich harvest of golden fruit, and then look at the work of the cowboy, branding, castrating, terrifying, butchering helpless animals; which is the more beautiful? Surely no one would say a corpse was a beautiful object. Picture it (after the axe has battered the skull, or the knife has found the heart, and the victim has at last ceased its dying groans and struggles), with its ghastly staring eyes, its blood-stained head or throat where the sharp steel pierced into the [Pg 47] quivering flesh; picture it when the body is opened emitting a sickening odour and the reeking entrails fall in a heap on the gore-splashed floor; picture this sight and ask whether it is not the epitome of ugliness, and in direct opposition to the most elementary sense of beauty.

Moreover, what effect has the work of a slayer of animals upon his personal character and refinement? Can anyone imagine a sensitive-minded, finely-wrought *æsthetic* nature doing anything else than revolt against the cold-blooded murdering of terrorised animals? It is significant that in some of the States of America butchers are not allowed to sit on a jury during a murder trial. Physiognomically the slaughterman carries his trade-mark legibly enough. The butcher does not usually exhibit those facial traits which distinguish a person who is naturally sympathetic and of an æsthetic temperament; on the contrary, the butcher's face and manner generally bear evidence of a life spent amid scenes of gory horror and violence; of a task which involves torture and death.

A plate of cereal served with fruit-juice pleases the eye and imagination, but a plate smeared with blood and laden with dead flesh becomes disgusting and repulsive the moment we consider it in that light. Cooking may disguise the appearance but cannot alter the reality of the decaying *corpse*; and to cook blood and give it another name (gravy) may be an artifice to please the palate, but it is blood, (blood that once coursed through [Pg 48] the body of a highly sensitive and nervous being), just the same. Surely a person whose olfactory nerves have not been blunted prefers the delicate aroma of ripe fruit to the sickly smell of mortifying flesh,—or fried eggs and bacon!

Notice how young children, whose taste is more or less unperverted, relish ripe fruits and nuts and clean tasting things in general. Man, before he has become thoroughly accustomed to an unnatural diet, before his taste has been perverted and he has acquired by habit a liking for unwholesome and unnatural food, has a healthy appetite for Nature's sun-cooked seeds and berries of all kinds. Now true refinement can only exist where the senses are uncorrupted by addiction to deleterious habits, and the nervous system by which the senses act will remain healthy only so long as it is built up by pure and natural foods; hence it is only while man is nourished by those foods desired by his unperverted appetite that he may be said to possess true refinement. Power of intellect has nothing whatever to do *necessarily* with the *æsthetic instinct*. A man may possess vast learning and yet be a boor. Refinement is not learnt as a boy learns algebra. Refinement comes from living a refined life, as good deeds come from a good man. The nearer we live according to Nature's plan, and in harmony with Her, the healthier we become physically and mentally. We do not look for refinement in the obese, red-faced, phlegmatic, gluttonous sensualists who often pass [Pg 49] as gentlemen because they possess money or rank, but in those who live simply, satisfying the simple requirements of the body, and finding happiness in a life of well-directed toil.

The taste of young children is often cited by vegetarians to demonstrate the liking of an unsophisticated palate, but the primitive instinct is not wholly atrophied in man. Before man became a tool-using animal, he must have depended for direction upon what is commonly termed instinct in the selection of a diet most suitable to his nature. No one can doubt, judging by the way undomesticated animals seek their food with unerring certainty as to its suitability, but that instinct is a trustworthy guide. Granting that man could, in a state of absolute savagery, and before he had discovered the use of fire or of tools, depend upon instinct alone, and in so doing live healthily, cannot *what yet remains* of instinct be of some value among civilized beings? Is not man, even now, in spite of his abused and corrupted senses, when he sees luscious fruits hanging within his reach, tempted to pluck them, and does he not eat them with relish? But when he sees the grazing ox, or the wallowing hog, do similar gustatory desires affect him? Or when he sees these animals lying dead, or when skinned and cut up in small pieces, does this same natural instinct stimulate him to steal and eat this food as it stimulates a boy to steal apples and nuts from an orchard and [Pg 50] eat them surreptitiously beneath the hedge or behind the haystack?

Very different is it with true carnivora. The gorge of a cat, for instance, will rise at the smell of a mouse, or a piece of raw flesh, but not at the aroma of fruit. If a man could take delight in pouncing upon a bird, tear its still living body apart with his teeth, sucking the warm blood, one might infer that Nature had provided him with carnivorous instinct, but the very *thought* of doing such a thing makes him shudder. On the other hand, a bunch of luscious grapes makes his 'mouth water,' and even in the absence of hunger he will eat fruit to gratify taste. A table spread with fruits and nuts and decorated with flowers is artistic; the same table laden with decaying flesh and blood, and maybe entrails, is not only inartistic—it is disgusting.

Those who believe in an all-wise Creator can hardly suppose He would have so made our body as to make it necessary daily to perform acts of violence that are an outrage to our sympathies, repulsive to our finer feelings, and brutalising and degrading in every detail. To possess fine feelings without the means to satisfy them is as bad as to possess hunger without a stomach. If it be necessary and a part of the Divine Wisdom that we should degrade ourselves to the level of beasts of prey, then the humanitarian sentiment and the æsthetic instinct are wrong and should be displaced by callousness, and the endeavour to cul [Pg 51] tivate a feeling of enjoyment in that which to all the organs of sense in a person of intelligence and religious feeling is ugly and repulsive. But no normally-minded person can think that this is so. It would be contrary to all the ethical and æsthetic teachings of every religion, and antagonistic to the feelings of all who have evolved to the possession of a conscience and the power to distinguish the beautiful from the base.

When one accustomed to an omnivorous diet adopts a vegetarian régime, a steadily growing refinement in taste and smell is experienced. Delicate and subtle flavours, hitherto unnoticed, especially if the habit of thorough mastication be practised, soon convince the neophyte that a vegetarian is by no means denied the pleasure of gustatory enjoyment. Further, not only are these senses better attuned and refined, but the mind also undergoes a similar exaltation. Thoreau, the transcendentalist, wrote: 'I believe that every man who has ever been earnest to preserve his higher or poetic faculties in the best condition, has been particularly inclined to abstain from animal food, and from much food of any kind.' [Pg 52]

V

ECONOMICAL CONSIDERATIONS

There is no doubt that the yield of land when utilized for pasturage is less than what it will produce in the hands of the agriculturist. In a thickly populated country, such as England, dependent under present conditions on foreign countries for a large proportion of her food supply, it is foolish, considering only the political aspects, to employ the land for raising unnecessary flesh-food, and so be compelled to apply to foreign markets for the first necessaries of life, when there is, without doubt, sufficient agricultural land in England to support the entire population on a vegetable regimen. As just said, a much larger population can be supported on a given acreage cultivated with vegetable produce than would be possible were the same land used for grazing cattle. Lieut. Powell quotes Prof. Francis Newman of University College, London, as declaring that—

100 acres devoted to sheep-raising will support 42 men: proportion 1. [Pg 53]

100 acres devoted to dairy-farming will support 53 men: proportion 1¼.

100 acres devoted to wheat will support 250 men: proportion 6.

100 acres devoted to potato will support 683 men: proportion 16.

To produce the same quantity of food yielded by an acre of land cultivated by the husbandman, three or four acres, or more, would be required as grazing land to raise cattle for flesh meat.

Another point to note is that agriculture affords employment to a very much larger number of men than cattle-raising; that is to say, a much larger number of men are required to raise a given amount of vegetable food than is required to raise the same amount of flesh food, and so, were the present common omnivorous customs to give place to vegetarianism, a very much more numerous peasantry would be required on the land. This would be physically, economically, morally, better for the nation. It is obvious that national health would be improved with a considerably larger proportion of hardy country yeomen. The percentage of poor and unemployed people in large cities would be reduced, their labor being required on the soil, where, being in more natural, salutary, harmonious surroundings the moral element would have better opportunity for development than when confined in the unhealthy, ugly, squalid surroundings of a city slum.

It is not generally known that there is often a decided *loss* of valuable food-material in feeding [Pg 54] animals for food, one authority stating that it takes nearly 4 lbs. of barley, which is a good wholesome food, to make 1 lb. of pork, a food that can hardly be considered safe to eat when we learn that tuberculosis was detected in 6,393 pigs in Berlin abattoirs in one year.

As to the comparative cost of a vegetarian and omnivorous diet, it is instructive to learn that it is proverbial in the Western States of America that a Chinaman can live and support his family in health and comfort on an allowance which to a meat-eating white man would be starvation. It is not to be denied that a vegetarian desirous of living to eat, and having no reason or desire to be economical, could spend money as extravagantly as a devotee of the flesh-pots having a similar disposition. But it is significant that the poor of most European countries are not vegetarians from choice but from necessity. Had they the means doubtless they would purchase meat, not because of any instinctive liking for it, but because of that almost universal trait of human character that causes men to desire to imitate their superiors, without, in most cases, any due consideration as to whether the supposed superiors are worthy of the genuflection they get. Were King George or Kaiser Wilhelm to become vegetarians and advocate the non-flesh diet, such an occurrence would do far more towards advancing the popularity of this diet than a thousand lectures from "mere" men of science. Carlyle was not far wrong when [Pg 55] he called men "clothes worshippers." The uneducated and poor imitate the educated and rich, not because they possess that attitude of mind which owes its existence to a very deep and subtle emotion and which is expressed in worship and veneration for power, whether it be power of body, power of rank, power of mind, or power of wealth. The poor among Western nations are vegetarians because they cannot afford to buy meat, and this is plain enough proof as to which dietary is the cheaper.

Perhaps a few straightforward facts on this point may prove interesting. An ordinary man, weighing 140 lbs. to 170 lbs., under ordinary conditions, at moderately active work, as an engineer, carpenter, etc., could live in comfort and maintain good health on a dietary providing daily 1 lb. bread (600 to 700 grs. protein); 8 ozs. potatoes (70 grs. protein); 3 ozs. rice, or barley, or macaroni, or maize meal, etc. (100 grs. protein); 4 ozs. dates, or figs, or prunes, or bananas, etc., and 2 ozs. shelled nuts (130 grs. protein); the cost of which need not exceed 10c. to 15c. per day; or in the case of one leading a more sedentary life, such as clerical work, these would be slightly reduced and the cost reduced to 8c. to 12c. per day. For one shilling per day, luxuries, such as nut butter, sweet-stuffs, and a variety of fruits and vegetables could be added. It is hardly necessary to point out that the housewife would be 'hard put to' to make ends meet [Pg 56] 'living well' on the ordinary diet at 25c. per head per day. The writer, weighing 140 lbs., who lives a moderately active life, enjoys good health, and whose tastes are simple, finds the cost of a cereal diet comes to 50c. to 75c. per week.

The political economist and reformer finds on investigation, that the adoption of vegetarianism would be a solution of many of the complex and baffling questions connected with the material prosperity of the nation. Here is a remedy for unemployment, drink, slums, disease, and many forms of vice; a remedy that is within the reach of everyone, and that costs only the relinquishing of a foolish prejudice and the adoption of a natural mode of living plus the effort to overcome a vicious habit and the denial of pleasure derived from the gratification of corrupted appetite. Nature will soon create a dislike for that which once was a pleasure, and in compensation will confer a wholesome and beneficent enjoyment in the partaking of pure and salutary foods. Whether or no the meat-eating nations will awake to these facts in time to save themselves from ruin and extinction remains to be seen. Meat-eating has grown side by side with disease in England during the past seventy years, but there are now, fortunately, some signs of abatement. The doctors, owing perhaps to some prescience in the air, some psychical foreboding, are recommending that less meat be eaten. But whatever the future has in store, there is nothing more certain than this—that [Pg 57] in the adoption of the vegetable regimen is to be found, if not a complete panacea, at least a partial remedy, for the political and social ills that our nation at the present time is afflicted with, and that those of us who would be true patriots are in duty bound to practise and preach vegetarianism wheresoever and whensoever we can. [Pg 58]

VI

THE EXCLUSION OF DAIRY PRODUCE

It is unfortunate that many flesh-abstainers who agree with the general trend of the foregoing arguments do not realise that these same arguments also apply to abstinence from those animal foods known as dairy produce. In considering this further aspect it is necessary for reasons already given, to place hygienic considerations first.

Is it reasonable to suppose that Nature ever intended the milk of the cow or the egg of the fowl for the use of man as food? Can anyone deny that Nature intended the cow's milk for the nourishment of her calf and the hen's egg for the propagation of her species? It is begging the question to say that the cow furnishes more milk than her calf requires, or that it does not injure the hen to steal her eggs. Besides, it is not true.

Regarding the dietetic value of milk and eggs, which is the question of first importance, are we correct in drawing the inference that as Nature did not intend these foods for man, therefore they are not suitable for him? As far as the chemical constituents of these foods are concerned, [Pg 59] it is true they contain compounds essential to the nourishment of the human body, and if this is going to be set up as an argument in favor of their consumption, let it be remembered that flesh food also contains compounds essential to nourishment. But the point is this: not what valuable nutritive compounds does any food-substance contain, but what value, *taking into consideration its total effects*, has the food in question as a wholesome article of diet?

It seems to be quite generally acknowledged by the medical profession that raw milk is a dangerous food on account of the fact that it is liable from various causes, sometimes inevitable, to contain impurities. Dr. Kellogg writes: Typhoid fever, cholera infantum, tuberculosis and tubercular consumption—three of the most deadly diseases known; it is very probable also, that diphtheria, scarlet fever and several other maladies are communicated through the medium of milk.... It is safe to say that very few people indeed are fully acquainted with the dangers to life and health which lurk in the milk supply.... The teeming millions of China, a country which contains nearly one-third of the entire population of the globe, are practically ignorant of this article of food. The high-class Hindoo regards milk as a loathsome and impure article of food, speaking of it with the greatest contempt as "cow-juice," doubtless because of his observations of the deleterious effect of the use of milk in its raw state. [Pg 60]

The germs of tuberculosis seem to be the most dangerous in milk, for they thrive and retain their vitality for many weeks, even in butter and cheese. An eminent German authority, Hirschberger, is said to have found 10 per cent of the cows in the vicinity of large cities to be affected by tuberculosis. Many other authorities might be quoted supporting the contention that a large percentage of cows are afflicted by this deadly disease. Other germs, quite as dangerous, find their way into milk in numerous ways. Excreta, clinging to the hairs of the udder, are frequently rubbed off into the pail by the action of the hand whilst milking. Under the most careful sanitary precautions it is impossible to obtain milk free from manure, from the ordinary germs of putrefaction to the most deadly microbes known to science. There is little doubt but that milk is one of the uncleanest and impurest of all foods.

Milk is constipating, and as constipation is one of the commonest complaints, a preventive may be found in abstinence from this food. As regards eggs, there is perhaps not so much to be said, although eggs so quickly undergo a change akin to putrefaction that unless eaten fresh they are unfit for food; moreover, (according to Dr. Haig) they contain a considerable amount of xanthins, and cannot, therefore, be considered a desirable food.

Dairy foods, we emphatically affirm, are not necessary to health. In the section dealing with [Pg 61] 'Physical Considerations' sufficient was said to prove the eminent value of an exclusive vegetable diet, and the reader is referred to that and the subsequent essay on Nutrition and Diet for proof that man can and should live without animal food of any kind. Such nutritive properties as are possessed by milk and eggs are abundantly found in the vegetable kingdom. The table of comparative values given, exhibits this quite plainly. That man can live a thoroughly healthy life upon vegetable foods alone there is ample evidence to prove, and there is good cause to believe that milk and eggs not only are quite unnecessary, but are foods unsuited to the human organism, and may be, and often are, the cause of disease. Of course, it is recognized that with scrupulous care this danger can be minimized to a great extent, but still it is always there, and as there is no reason why we should consume such foods, it is not foolish to continue to do so?

But this is not all. It is quite as impossible to consume dairy produce without slaughter as it is to eat flesh without slaughter. There are probably as many bulls born as cows. One bull for breeding purposes suffices for many cows and lives for many years, so what is to be done with the bull calves if our humanitarian scruples debar us from providing a vocation for the butcher? The country would soon be overrun with vast herds of wild animals and the whole populace would have to take to arms for self-preservation. So [Pg 62] it comes to the same thing. If we did not breed these animals for their flesh, or milk, or eggs, or labour, we should have no use for them, and so should breed them no longer, and they would quickly become extinct. The wild goat and sheep and the feathered life might survive indefinitely in mountainous districts, but large animals that are not domesticated, or bred for slaughter, soon disappear before the approach of civilisation. The Irish elk is extinct, and the buffalo of North America has been wiped out during quite recent years. If leather became more expensive (much of it is derived from horse hide) manufacturers of leather substitutes would have a better market than they have at present. [Pg 63]

VI

CONCLUSION

'However much thou art read in theory, if thou hast no practice thou art ignorant,' says the Persian poet Sa'di. 'Conviction, were it never so excellent, is worthless until it converts itself into Conduct. Nay, properly, Conviction is not possible till then,' says Herr Teufelsdrockh. It is never too late to be virtuous. It is right that we should look before we leap, but it is gross misconduct to neglect duty to conform to the consuetudes of the hour. We must endeavour in practical life to carry out to the best of our ability our philosophical and ethical convictions, for any lapse in such endeavour is what constitutes immorality. We must live consistently with theory so long as our chief purpose in life is advanced by so doing, but we must be inconsistent when by antinomianism we better forward this purpose. To illustrate: All morally-minded people desire to serve as a force working for the happiness of the race. We are convinced that the slaughter of animals for food is needless, and that it entails much physical and mental suffering among men [Pg 64] and animals and is therefore immoral. Knowing this we should exert our best efforts to counteract the wrong, firstly, by regulating our own conduct so as not to take either an active or passive part in this needless massacre of sub-human life, and secondly, by making those facts widely known which show the necessity for food reform.

Now to go to the ultimate extreme as regards our own conduct we should make no use of such things as leather, bone, catgut, etc. We should not even so much as attend a concert where the players use catgut strings, for however far distantly related cause and effect may be, the fact remains that the more the demand, no matter how small, the more the supply. We should not even be guilty of accosting a friend from over the way lest in consequence he take more steps than otherwise he would do, thus wearing out more shoe-leather. He who would practise such absurd sansculottism as this would have to resort to the severest seclusion, and plainly enough we cannot approve of such fanaticism. By turning antinomian when necessary and staying amongst our fellows, making known our views according to our ability and opportunity, we shall be doing more towards establishing the proper relation between man and sub-man than by turning cenobite and refusing all intercourse and association with our fellows. Let us do small wrong that we may accomplish great good. Let us practise our creed so far as to abstain from the eating of [Pg 65] animal food, and from the use of furs, feathers, seal and fox skins, and similar ornaments, to obtain which necessitates the violation of our fundamental principles. With regard to leather, this material is, under present conditions, a 'by-product.' The hides of animals slaughtered for their flesh are made into leather, and it is not censurable in a vegetarian to use this article in the absence of a suitable substitute when he knows that by so doing he is not asking an animal's life, nor a fellow-being to degrade his character by taking it. There is a substitute for leather now on the market, and it is hoped that it may soon be in demand, for even a leather-tanner's work is not exactly an ideal occupation.

Looking at the question of conviction and consistency in this way, there are conceivable circumstances when the staunchest vegetarian may even turn kreophagist. As to how far it is permissible to depart from the strictest adherence to the principles of vegetarianism that have been laid down, the individual must trust his own conscience to determine; but we can confidently affirm that the eating of animal flesh is unnecessary and immoral and retards development in the direction which the finest minds of the race hold to be good; and that the only time when it would not be wrong to feed upon such food would be when, owing to misfortunes such as shipwreck, war, famine, etc., starvation can only be kept at bay by the sacrifice of animal life. In such a case, man, con [Pg 66] sidering his own life the more valuable, must resort to the unnatural practice of flesh-eating.

The reformer may have, indeed must have, to pay a price, and sometimes a big one, for the privilege, the greatest of all privileges, of educating his fellows to a realisation of their errors, to a realisation of a better and nobler view of life than they have hitherto known. Seldom do men who carve out a way for themselves, casting aside the conventional prejudices of their day, and daring to proclaim, and live up to, the truth they see, meet with the esteem and respect due to them; but this should not, and, if they are sincere and courageous, does not, deter them from announcing their message and caring for the personal discomfort it causes. It is such as these that the world has to thank for its progress.

It often happens that the reformer reaps not the benefit of the reform he introduces. Men are slow to perceive and strangely slow to act, yet he who has genuine affection for his fellows, and whose desire for the betterment of humanity is no mere sentimental pseudo-religiosity, bears bravely the disappointment he is sure to experience, and with undaunted heart urges the cause that, as he sees it, stands for the enlightenment and happiness of man. The vegetarian in the West (Europe, America, etc.) is often ridiculed and spoken of by appellations neither complimentary nor kind, but this should deter no honorable man or woman from entering the ranks of [Pg 67] the vegetarian movement as soon as he or she perceives the moral obligation to do so. It may be hard, perhaps impossible, to convert others to the same views, but the vegetarian is not hindered from living his own life according to the dictates of his conscience. 'He who conquers others is strong, but the man who conquers himself is mighty,' wrote Laotze in the *Tao Teh Ch'ing*, or 'The Simple Way.'

When we call to mind some heroic character—a Socrates, a Regulus, a Savonarola—the petty sacrifices our duties entail seem trivial indeed. We do well to remember that it is only by obedience to the highest dictates of our own hearts and minds that we may obtain true happiness. It is only by living in harmony with all living creatures that nobility and purity of life are attainable. As we obey the immediate vision, so do we become able to see yet richer visions: but the *strength of the vision is ours only as we obey its high demands.*
[Pg 68]
[Pg 69]

[Pg 70]

NUTRITION AND DIET

1

THE SCIENCE OF NUTRITION

The importance of some general knowledge of the principles of nutrition and the nutritive values of foods is not generally realised. Ignorance on such a matter is not usually looked upon as a disgrace, but, on the contrary, it would be commonly thought far more reprehensible to lack the ability to conjugate the verb 'to be' than to lack a knowledge of the chemical properties of the food we eat, and the suitability of it to our organism. Yet the latter bears direct and intimate relation to man's physical, mental, and moral well-being, while the former is but a 'sapless, heartless thistle for pedantic chaffinches,' as Jean Paul would say.

The human body is the most complicated machine conceivable, and as it is absurd to suppose that any tyro can take charge of so comparatively simple a piece of mechanism as a locomotive, how much more absurd is it to suppose the human body can be kept in fit condition, and worked satisfactorily, without at least some, if only slight, knowledge of the nature of its constitution, and an [Pg 71] understanding of the means to satisfy its requirements? Only by study and observation comes the knowledge of how best to supply the required material which, by its oxidation in the body, repairs waste, gives warmth and produces energy.

Considering, then, that the majority of people are entirely ignorant both of the chemical constitution of the body, and the physiological relationship between the body and food, it is not surprising to observe that in respect to this question of caring for the body, making it grow and work and think, many come to grief, having breakdowns which are called by various big-sounding names. Indeed, to the student of dietetics, the surprise is that the body is so well able to withstand the abuse it receives.

It has already been explained in the previous essay how essential it is if we live in an artificial environment and depart from primitive habits, thereby losing natural instincts such as guide the wild animals, that we should study diet. No more need be said on this point. It may not be necessary that we should have some general knowledge of fundamental principles, and learn how to apply them with reasonable precision.

The chemical constitution of the human body is made up of a large variety of elements and compounds. From fifteen to twenty elements are found in it, chief among which are oxygen, hydrogen, carbon, nitrogen, calcium, phosphorus, sodium, and sulphur. The most important compounds [Pg 72] are protein, hydrocarbons, carbohydrates, organic mineral matter, and water. The food which nourishes the body is composed of the same elements and compounds.

Food serves two purposes,—it builds and repairs the body tissues, and it generates vital heat and energy, burning food as fuel. Protein and mineral matter serve the first purpose, and hydrocarbons (fats) and carbohydrates (sugars and starches) the second, although, if too much protein be assimilated it will be burnt as fuel, (but it is bad fuel as will be mentioned later), and if too much fat is consumed it will be stored away in the body as reserve supply. Most food contains some protein, fat, carbohydrates, mineral matter, and water, but the proportion varies very considerably in different foods.

Water is the most abundant compound in the body, forming on an average, over sixty per cent. of the body by weight. It cannot be burnt, but is a component part of all the tissues and is therefore an exceedingly, important food. Mineral matter forms approximately five or six per cent. of the body by weight. Phosphate of lime (calcium phosphate), builds bone; and many compounds of potassium, sodium, magnesium and iron are present in the body and are necessary nutrients. Under the term protein are included the principal nitrogenous compounds which make bone, muscle and other material. It forms about 15 per cent. of the body by weight, and, as men

[Pg 73] tioned above, is burnt as fuel for generating heat and energy. Carbohydrates form but a small proportion of the body-tissue, less than one per cent. Starches, sugars, and the fibre of plants, or cellulose, are included under this term. They serve the same purpose as fat.

All dietitians are agreed that protein is the essential combined in food. Deprivation of it quickly produces a starved physical condition. The actual quantity required cannot be determined with perfect accuracy, although estimates can be made approximately correct. The importance of the other nutrient compounds is but secondary. But the system must have all the nutrient compounds in correct proportions if it is to be maintained in perfect health. These proportions differ slightly according to the individual's physical constitution, temperament and occupation.

Food replenishes waste caused by the continual wear and tear incidental to daily life: the wear and tear of the muscles in all physical exertion, of the brain in thinking, of the internal organs in the digestion of food, in all the intricate processes of metabolism, in the excretion of waste matter, and the secretion of vital fluids, etc. The ideal diet is one which replenishes waste with the smallest amount of suitable material, so that the system is kept in its normal condition of health at a minimum of expense of energy. The value, therefore, of some general knowledge of the chemical constituents of food is obvious. The diet [Pg 74] must be properly balanced, that is, the food eaten must provide the nutrients the body requires, and not contain an excess of one element or a deficiency of another. It is impossible to substitute protein for fat, or *vice versa*, and get the same physiological result, although the human organism is wonderfully tolerant of abuse, and remarkably ingenious in its ability to adapt itself to abnormal conditions.

It has been argued that it is essentially necessary for a well-balanced dietary that the variety of food be large, or if the variety is to be for any reason restricted, it must be chosen with great discretion. Dietetic authorities are not agreed as to whether the variety should be large or small, but there is a concensus of opinion that, be it large or small, it should be selected with a view to supplying the proper nutrients in proper proportions. The arguments, so far as the writer understands them, for and against a large variety of foods, are as follows: —

If the variety be large there is a temptation to over-feed. Appetite does not need to be goaded by tasty dishes; it does not need to be goaded at all. We should eat when hungry and until replenished; but to eat when not hungry in order to gratify a merely sensual appetite, to have dishes so spiced and concocted as to stimulate a jaded appetite by novelty of taste, is harmful to an extent but seldom realised. Hence the advisability, at least in the case of persons who have [Pg 75] not attained self-mastery over sensual desire, of having little variety, for then, when the system is replenished, overfeeding is less likely to occur.

In this connection it should be remembered that in some parts of the world the poor, although possessing great strength and excellent health, live upon, and apparently relish, a dietary limited mostly to black bread and garlics, while among ourselves an ordinary person eats as many as fifty different foods in one day. [3]

On the other hand, a too monotonous dietary, especially where people are accustomed to a large variety of mixed foods, fails to give the gustatory pleasure necessary for a healthy secretion of the digestive juices, and so may quite possibly result in indigestion. It is a matter of common observation that we are better able to digest food which we enjoy than that which we dislike, and as we live not upon what we eat, but upon what we digest, the importance of enjoying the food eaten is obvious.

Also as few people know anything about the nutritive value of foods, they stand a better chance, if they eat a large variety, of procuring the required quantity of different nutrients than when restricted to a very limited dietary, because, [Pg 76] if the dietary be very limited they might by accident choose as their mainstay some food that was badly balanced in the different nutrients, perhaps wholly lacking in protein. It is lamentable that there is such ignorance on such an all-important subject. However, we have to consider things as they are and not as they ought to be.

Perhaps the best way is to have different food at different meals, without indulging in many varieties at one meal. Thus taste can be satisfied, while the temptation to eat merely for the sake of eating is less likely to arise.

It might be mentioned, in passing, that in the opinion of the best modern authorities the average person eats far more than he needs, and that this excess inevitably results in pathological conditions. Voit's estimate of what food the average person requires daily was based upon observation of what people *do* eat, not upon what they *should* eat. Obviously such an estimate is valueless. As well argue that an ounce of tobacco daily is what an ordinary person should smoke because it is the amount which the average smoker consumes.

A vegetarian needs only to consider the amount of protein necessary, and obtained from the food eaten. The other nutrients will be supplied in proportions correct enough to satisfy the body requirements under normal conditions of health. The only thing to take note of is that more fat and carbohydrates are needed in cold weather than hot, the body requiring more fuel for warmth. [Pg 77] But even this is not essential: the essential thing is to have the required amount of protein. In passing, it is interesting to observe the following: the fact that in a mixed fruitarian diet the proportion of the nutrient compounds is such as to satisfy natural requirements is another proof of the suitability of the vegetable regimen to the human organism. It is a provision of Nature that those foods man's digestive organs are constructed to assimilate with facility, and man's organs of taste, smell, and perception best prefer, are those foods containing chemical compounds in proportions best suited to nourish his body.

One of the many reasons why flesh-eating is deleterious is that flesh is an ill-balanced food, containing, as it does, considerable protein and fat, but no carbohydrates or neutralising salts whatever. As the body requires three to four times more carbohydrates than protein, and protein cannot be properly assimilated without organic minerals, it is seen that with the customary 'bread, meat and boiled potatoes' diet, this proportion is not obtained. Prof. Chittenden holds the opinion that the majority of people partake greatly in excess of food rich in protein.

No hard and fast rule can be laid down to different persons require different foods and foods and amounts at different times under different

Transcriber's note: It is regretted that a line has been missed out by the typesetter.

regulate the amount, or proper proportions, of food material for a well-balanced dietary, as amounts, and the same person requires different [Pg 78] ferent conditions. Professor W. O. Atwater, an American, makes the following statement: 'As the habits and conditions of individuals differ, so, too, their needs for nourishment differ, and their food should be adapted to their particular requirements. It has been estimated that an average man at moderately active labor, like a carpenter, or mason, should have (daily) about 115 grams (1750 grains) or 0.25 pound of available protein, and sufficient fuel ingredients in addition to make the fuel value of the whole diet 3,400 calories; while a man at sedentary employment would be well nourished with 92 grams (1400 grains) or 0.20 pound of available protein, and enough fat and carbohydrates in addition to yield 2,700 calories of energy. The demands are, however, variable, increasing and decreasing with increase and decrease of muscular work, or as other needs of the person change. Each person, too, should learn by experience what kinds of food yield him nourishment with the least discomfort, and should avoid those which do not "agree" with him.'

It has been stated that unless the body is supplied with protein, hunger will be felt, no matter if the stomach be over-loaded with non-nitrogenous food. If a hungry man ate heartily of *only* such foods as fresh fruit and green vegetables he might soon experience a feeling of fulness, but his hunger would not be appeased. Nature asks for protein, and hunger will continue so long as this want remains unsatisfied. Similarly [Pg 79] as food is the first necessity of life, so is protein the first necessity in food. If a person were deprived of protein starvation must inevitably ensue.

Were we (by 'we' is meant the generality of people in this country), to weigh out our food supply, for, say a week, we should soon realise what a large reduction from the usual quantity of food consumed would have to be made, and instead of eating, as is customary, without an appetite, hunger might perhaps once a day make itself felt. There is little doubt but that the health of most people would be vastly improved if food were only eaten when genuine hunger was felt, and the dietary chosen were well balanced, *i.e.*, the proportions of protein, fat, carbohydrates and salts being about 3, 2, 9, 2-3. As aforesaid, the mixed vegetarian dietary is, in general, well-balanced.

While speaking about too much food, it may be pointed out that the function of appetite is to inform us that the body is in need of nutriment. The appetite was intended by Nature for this purpose, yet how few people wait upon appetite! The generality of people eat by time, custom, habit, and sensual desire; not by appetite at all. If we eat when not hungry, and drink when not thirsty, we are doing the body no good but positive harm. The organs of digestion are given work that is unnecessary, thus detracting from the vital force of the body, for there is only a limited amount of potential energy, and if some of this [Pg 80] is spent unnecessarily in working the internal organs, it follows that there is less energy for working the muscles or the brain. So that an individual who habitually overfeeds becomes, after a time, easily tired, physically lazy, weak, perhaps if temperamentally predisposed, nervous and hypochondriacal. Moreover, over-eating not only adds to the general wear and tear, thus probably shortening life, but may even result in positive disease, as well as many minor complaints such as constipation, dyspepsia, flatulency, obesity, skin troubles, rheumatism, lethargy, etc.

Just as there is danger in eating too much, so there is much harm done by drinking too much. The evil of stimulating drinks will be spoken of later; at present reference is made only to water and harmless concoctions such as lime-juice, unfermented wines, etc. To drink when thirsty is right and natural; it shows that the blood is concentrated and is in want of fluid. But to drink merely for the pleasure of drinking, or to carry out some insane theory like that of 'washing out' the system is positively dangerous. The human body is not a dirty barrel needing swilling out with a hose-pipe. It is a most delicate piece of mechanism, so delicate that the abuse of any of its parts tends to throw the entire system out of order. It is the function of the blood to remove all the waste products from the tissues and to supply the fresh material to take the place of that which has been removed. Swilling the system out [Pg 81] with liquid does not in any way accelerate or aid the process, but, on the contrary, retards and impedes it. It dilutes the blood, thus creating an abnormal condition in the circulatory system, and may raise the pressure of blood and dilate the heart. Also it dilutes the secretions which will therefore 'act slowly and inefficiently, and more or less fermentation and putrefaction will meanwhile be going on in the food masses, resulting in the formation of gases, acids, and decomposition products.'

Eating and drinking too much are largely the outcome of sensuality. To see a man eat sensually is to know how great a sensualist he is. Sensualism is a vice which manifests itself in many forms. Poverty has its blessings. It compels abstinence from rich and expensive foods and provides no means for surfeit. Epicurus was not a glutton. Socrates lived on bread and water, as did Sir Isaac Newton. Mental culture is not fostered by gluttony, but gluttony is indulged in at the expense of mental culture. The majority of the world's greatest men have led comparatively simple lives, and have regarded the body as a temple to be kept pure and holy.

We have now to consider (*a*) what to eat, (*b*) when to eat, (*c*) how to eat. First, then, we will consider the nutritive properties of the common food-stuffs. [Pg 82]

II

WHAT TO EAT

Among the foods rich in protein are the legumes, the cereals, and nuts. Those low in protein are fresh fruits, green vegetables, and roots. Fat is chiefly found in nuts, olives, and certain pulses, particularly the peanut; and carbohydrates in cereals, pulses, and many roots. Fruit and green vegetables consist mostly of water and organic mineral compounds, and in the case of the most juicy varieties may be regarded more as drink than food. We have, then, six distinct classes of food — the pulses, cereals, nuts, fruits, green vegetables, and roots. Let us briefly consider the nutritive value of each.

Pulse foods usually form an important item in a vegetarian dietary. They are very rich in their nutritive properties, and even before matured are equal or superior in value to any other green vegetable. 'The ripened seed shows by analysis a very remarkable contrast to most of the matured foods, as the potato and other tubers, and even to the best cereals, as wheat. This superiority lies in the large amount of nitrogen in the form [Pg 83] of protein that they contain.' Peas, beans, and lentils should be eaten very moderately, being highly concentrated foods. The removal of the skins from peas and beans, also of the germs of beans, by parboiling, is recommended, as they are then more easily digested and less liable to 'disagree.' These foods, it is interesting to know are used extensively by the vegetarian nations. The Mongol procures his supply of protein chiefly from the Soya bean from which he makes different preparations of bean cheese and sauce. It is said that the poorer classes of Spaniards and the Bedouins rely on a porridge of lentils for their mainstay. In India and China where rice is the staple food, beans are eaten to provide the necessary nitrogenous matter, as rice alone is considered deficient in protein.

With regard to the pulse foods, Dr. Haig, in his works on uric acid, states that, containing as they do considerable xanthin, an exceedingly harmful poison, they are not to be commended as healthful articles of diet. He states that he has found the pulses to contain even more xanthin than many kinds of flesh-meat, and as it is this poison in flesh that causes him to so strongly condemn the eating of meat, he naturally condemns the eating of any foods in which this poison exists in any considerable quantity. He writes: 'So far as I know the "vegetarians" of this country are decidedly superior in endurance to those feeding on animal tissues, who might [Pg 84] otherwise be expected to equal them; but these "vegetarians" would be still better if they not only ruled out animal flesh, but also eggs, the pulses (peas, beans, lentils and peanuts), eschew nuts, asparagus, and mushrooms, as well as tea, coffee and cocoa, all of which contain a large amount of uric acid, or substances physiologically equivalent to it.'

Dr. Haig attributes many diseases and complaints to the presence of uric acid in the blood and its deposits in the tissues: 'Uric acid diseases fall chiefly in two groups: (a) The arthritic group, comprising gout, rheumatism, and similar affections of many fibrous tissues throughout the body; (b) the circulation group including headache, epilepsy, mental depression, anæmia, Bright's disease, etc.' Speaking with regard to rheumatism met with among the vegetarian natives of India, Dr. Haig writes: 'I believe it will appear, on investigation, that in those parts of India where rice and fresh vegetables form the staple foods, not only rheumatism, but uric acid diseases generally are little known, whereas in those parts where pulses are largely consumed, they are common—almost universal.'

The cereals constitute the mainstay of vegetarians all the world over, and although not superior to nuts, must be considered an exceedingly valuable, and, in some cases, essential food material. They differ considerably in their nu [Pg 85] tritive properties, so it is necessary to examine the worth of each separately.

Wheat, though not universally the most extensively used of the cereals, is the most popular and best known cereal in this country. It has been cultivated for ages and has been used by nearly all peoples. It is customary to grind the berries into a fine meal which is mixed with water and baked. There are various opinions about the comparative value of white and whole-wheat flour. There is no doubt but that the whole-wheat flour containing, as it does, more woody fibre than the white, has a tendency to increase the peristaltic action of the intestines, and thus is valuable for persons troubled with constipation. [4] From a large number of analyses it has been determined that entire wheat flour contains about 2.4 per cent. more protein than white flour (all grades), yet experiments have demonstrated that the *available* protein is less in entire wheat-flour than in white flour. [5] This is probably due to the fact that the protein which is enclosed in the bran cannot be easily assimilated, as the digestive organs are unable to break up the outer walls of woody fibre and extract the nitrogenous matter they contain. On the other hand whole-wheat flour contains con [Pg 86] siderably more valuable and available mineral matter than does white flour. The two outer layers contain compounds of phosphorus, lime, iron, and soda. Analyses by Atwater show entire-wheat flour to contain twice as much mineral matter as white flour. It is affirmed by Broadbent and others, that this mineral matter is exceedingly valuable both as a nutrient, and because of its neutralising effect upon proteid wastes, and that it is because of this that flour made from the entire-wheat berry has very superior food value to that made from the berry minus the outer cuticles. Many dietetists look upon whole-wheat bread as one of the most salutary of all foods and strongly advise its use in place of white bread. A well-known doctor states that he has known it a cure for many diseases, and thinks that many nervous complaints due to 'saline starvation' can be cured by substituting whole-meal for white bread.

But in opposition to these views Dr. Haig thinks that as the outer brown husk of all cereals contains some xanthin, it should on this account be removed. He therefore recommends white flour, (not superfine, but cheap-grade), in place of the entire-wheat. Others, however, are of the opinion that the amount of xanthin present in the bran is so small as not to be considered, especially when, by the removal of the xanthin, valuable mineral matter is also removed.

Of course, it is difficult for a layman to form an opinion when experts differ. Perhaps the best [Pg 87] thing to do is to use whole-wheat bread if there is any tendency to constipation. If not, then choose that which is the more palatable, or change from one to the other as inclination dictates. This adds to variety, and as digestion is better when the food is better relished, no doubt, in this case, that which pleases the taste best is the best to eat. At least, we can hold this view tentatively for the present.

Wheat flour (entire), ranks the highest of all the cereals in protein, excepting oatmeal, averaging 13 per cent. In fat it exceeds rice and rye, is equal with barley and maize, but considerably below oatmeal: averaging about 1.9 per cent. In carbohydrates it averages about seventy-two per cent., all the cereals being very much alike in quantity of these nutrients. It is a well-balanced food, as indeed, all cereals are, and is palatable prepared in a variety of ways, although, made into unleavened, unsalted bread, the sweet, nutty flavour of the berry itself is best preserved.

Oatmeal is not extensively used, comparatively speaking, although it has an excellent reputation. It is decidedly the richest cereal in protein and fat, especially fat, and this is probably why people living in cold climates find it such a sustaining food. In protein it averages 16.1 per cent.: in fat 7.2 per cent. It is very commonly used as porridge. When well cooked, that is to say, for several hours, this is a good way to prepare it, but a better is to eat it dry in the form [Pg 88] of unsweetened oatcakes, scones, etc., these being more easily digested because necessitating thorough mastication. The above remarks regarding the removal of the bran from wheat-flour are precisely as applicable to oatmeal, as well as rye, so no more need be said on that point.

Rye flour is not unlike wheat, and is used more extensively than wheat in many parts of Europe. It has 2 per cent. less protein than wheat and its gluten is darker in colour and less elastic and so does not make as light a loaf; but this does not detract from its nutritive value at all. Being more easily cultivated than wheat, especially in cold countries, it is cheaper and therefore more of a poor man's food.

Indian corn, or maize, or Turkish wheat, is one of the finest of cereals. It is used extensively in America, North and South, in parts of the Orient, in Italy, the Balkans, Servia, and elsewhere. It is used as a green vegetable and when fully matured is ground into meal and made into bread, porridge, biscuits, Johnny-cake, etc., etc. Corn compared to wheat is rich in fat, but in protein wheat is the richer by about 3 per cent. Sugar corn, cooked and canned, is sold in England by food-reform dealers. It is perhaps the most tasty of all the cereals.

Rice is the staple of the Orientals. The practice of removing the dark inner skin in order to give the uncooked grain a white and polished appearance, is not only an expensive operation, [Pg 89] but a very foolish one, for it detracts largely from the nutritive value of the food, as considerable protein and other valuable matter is removed along with the bran. We are told that the Burmese and Japanese and other nations who use rice as their principal food-stuff, use the entire grain. As compared to undressed rice, the ordinary, or polished rice is deficient 3 per cent. of protein; 6 per cent. of fat; 5 per cent. of mineral matter. 'Once milled' rice can be procured in this country, but has to be specially asked for. Rice is not nearly so nitrogenous as wheat, but is equal to it in fuel value, this being due to the large amount of starch it contains. It is an excellent food, being easily digested and easily prepared.

Millet, buckwheat, wild rice, sesame, and Kaffir corn, are cereals little known in this country, although where they are raised they are largely used by the natives. However, we need not trouble to consider their food value as they are not easily procurable either in Europe or America.

Nuts are perhaps the best of all foods. There is no doubt but that man in his original wild state lived on nuts and berries and perhaps roots. Nuts are rich in protein and fat. They are a concentrated food, very palatable, gently laxative, require no preparation but shelling, keep well, are easily portable, and are, in every sense, an ideal food. They have a name for being indigestible, but this may be due to errors in eating, not to the nuts. If we eat nuts, as is often done, after [Pg 90] having loaded the stomach with a large dinner, the work of digesting them is rendered very difficult, for the digestive apparatus tires itself disposing of the meal just previously eaten. Most things are indigestible eaten under such conditions. Nuts should be looked upon as the essential part of the meal and should be eaten first; bread, salad stuffs and fruit help to supply bulk and can follow as dessert if desired. Another cause of nuts not being easily digested is insufficient mastication. They are hard, solid food, and should be thoroughly chewed and insalivated before being swallowed. If the teeth are not good, nuts may be grated in an ordinary nut-mill, and then, if eaten slowly and sparingly, will generally be found to digest. Of course with a weak digestion nuts may have to be avoided, or used in very small quantities until the digestion is strengthened; but with a normal, healthy person, nuts are a perfect food and can be eaten all the year round. Perhaps it is best not to eat a large quantity at once, but to spread the day's supply over four or five light meals. With some, however, two meals a day seems to work well.

Pine kernels are very suitable for those who have any difficulty in masticating or digesting the harder nuts, such as the brazil, filbert, etc. They are quite soft and can easily be ground into a soft paste with a pestil and mortar, making delicious butter. They vary considerably in nitrogenous matter, averaging about 25 per cent. and [Pg 91] are very rich in fat, averaging about 50 per cent. Chestnuts are used largely by the peasants of Italy. They are best cooked until quite soft when they are easily digested. Chestnut meal is obtainable, and when combined with wheatmeal is useful for making biscuits and breadstuffs. Protein in chestnuts averages 10 per cent. Walnuts, Hazelnuts, Filberts, Brazils, Pecans, Hickory nuts, Beechnuts, Butternuts, Pistachio nuts and Almonds average 16 per cent. protein; 52 per cent. fat; 20 per cent. carbohydrates; 2 per cent. mineral salts. As each possesses a distinct flavour, one can live on nuts alone and still enjoy the pleasure of variety. A man weighing 140 lbs. would, at moderately active labour, require, to live on almonds alone—11 ozs. per day. 10 ozs. of nuts per day together with some fresh fruit or green salad in summer, and in winter, some roots, as potato, carrot, or beetroot, would furnish an ideal diet for one whose taste was simple enough to relish it.

Fruits are best left alone in winter. They are generally acid, and the system is better without very acid foods in the cold weather. But fruits are health-giving foods in warm and hot weather, and living under natural, primitive conditions, this is the only time of the year we should have them, for Nature only provides fruit during the months of summer. The fraction of protein fruit contains, 1 per cent. or less, is too small to be of any [Pg 92] account. The nutritive value of fruits consists in their mineral salts, grape-sugar and water.

Much the same applies to green vegetables. In cooking vegetables care should be taken that the water they are cooked in is not thrown away as it contains nearly all the nutrient properties of the vegetable; that is to say, the various salts in the vegetable become dissolved in the water they are boiled in. This water can be used for soup if desired, or evaporated, and with flour added to thicken, served as sauce to the vegetable. Potatoes are a salutary food, especially in winter. They contain alkalies which help to lessen the accumulation of uric acid. They should be cooked with skins on: 16 grains per lb. more of valuable potash salts are thus obtained than when peeled and boiled in the ordinary way. The ideal method, however, of taking most vegetables is in the form of uncooked salads, for in these the health-giving, vitalising elements remain unaltered.

If man is to be regarded, as many scientists regard him, as a frugivore, constitutionally adapted and suited to a nut-fruit diet, then to regain our lost original taste and acquire a liking for such simple foods should be our aim. It may be difficult, if not impossible, to make a sudden change after having lived for many years upon the complex concoctions of the chef's art, for the system resents sudden changes, but with proper care, changing discreetly, one can generally attain [Pg 93] a desired end, especially when it involves the replacing of a bad habit by a good one.

In the recipes that follow no mention is made of condiments, *i.e.*, pepper, salt, mustard, spice, *et hoc genus omni*. Condiments are not foods in any sense whatever, and the effect upon the system of 'seasoning' foods with these artificial aids to appetite, is always deleterious, none the less because it may at the time be imperceptible, and may eventually result in disease. Dr. Kellogg writes: 'By contact, they irritate the mucous membrane, causing congestion and diminished secretion of gastric juice when taken in any but quite small quantities. When taken in quantities so small as to occasion no considerable irritation of the mucous membrane, condiments may still work injury by their stimulating effects, when long continued.... Experimental evidence shows that human beings, as well as animals of all classes, live and thrive as well without salt as with it, other conditions being equally favorable. This statement is made with a full knowledge of counter arguments and experiments, but with abundant testimony to support the position taken.... All condiments hinder natural digestion.'

Condiments, together with such things as pickles, vinegar, alcohol, tea, coffee, cocoa, tobacco, opium, are all injurious, and undoubtedly are the cause of an almost innumerable number of minor, and, in some cases, serious, complaints. Theine, caffeine, and theobromine, all stimulant drugs, are [Pg 94] present in tea, coffee, and cocoa, respectively. Tea also contains tannin, a substance which is said to seriously impair digestion.

Alcohol, tea, coffee, etc., are stimulants. Stimulants do not produce force and should never be mistaken for food. They are undoubtedly injurious, as they are the cause, among other evils, of *loss* of force. They cause an abnormal metabolism which ultimately weakens and exhausts the whole system. While these internal activities are taking place, artificial feelings of well-being, or, at least, agreeable sensations, are produced, which are unfortunately mistaken for signs of benefit. Speaking of alcohol Dr. Haig writes: 'It introduces no albumen or force, it merely affects circulation, nutrition, and the metabolism of the albumens already in the body, and this call on the resources of the body is invariably followed by a corresponding depression or economy in the future.... It has been truly said that the man who relies upon stimulants for strength is lost, for he is drawing upon a reserve fund, which is not completely replaced, and physiological bankruptcy must inevitably ensue. This is what the stimulants such as tea, coffee, alcohol, tobacco, opium and cocaine do for those who trust in them.[1]

He who desires to enjoy life desires to possess good physical health, for a healthy body is almost essential to a happy life; and he who desires to live healthily does not abuse his body with poisonous drugs. It may require courage to reform, [Pg 95] but he who reforms in this direction has the satisfaction of knowing that his good health will probably some day excite the envy of his critics.

The chemical composition of all the common food materials can be seen from tables of analyses. It would be to the advantage of everyone to spend a little time examining these tables. It is not a difficult matter, and the trouble to calculate the quantity of protein in a given quantity of food, when once the *modus operandi* is understood, is trifling. As it has not unwisely been suggested, if people would give, say, one-hundredth the time and attention to studying the needs of the body and how to satisfy them as they give to dress and amusement, there is little doubt that there would be more happiness in the world.

The amount of protein in any particular prepared food is arrived at in the following manner: In the first place those ingredients containing a noticeable amount of protein are carefully weighed. Food tables are then consulted to discover the protein percentage. Suppose, for instance, the only ingredient having a noticeable quantity of protein is rice, and 1 lb. is used. The table is consulted and shows rice to contain eight per cent. protein. In 1 lb. avoirdupois there are 7,000 grains; eight per cent. of 7,000 is $70.00 \times 8 = 560$ grains. Therefore, in the dish prepared there are 560 grains of protein. It is as well after cooking to weight the entree or pudding and divide the number of ounces it weighs into 560, [Pg 96] thus obtaining the number of grains per ounce. Weighing out food at meals is only necessary at first, say for the first week or so. Having decided about how many grains of protein to have daily, and knowing how many grains per ounce the food contains, the eye will soon get trained to estimate the quantity needed. It is not necessary to be exact; a rough approximation is all that is needed, so as to be sure that the system is getting somewhere near the required amount of nutriment, and not suffering from either a large excess or deficiency of protein. [Pg 97]

III

WHEN TO EAT

The question of when to eat is of some importance. The Orientals eat fewer meals than we do, and in their abstemiousness they set us an example we should do well to follow. Sufficient has already been said to show that it is a mistake to imagine a great deal of food gives great strength. When we eat frequently, and especially when we 'live well,' that is, are accustomed to a large variety of food, we are tempted to eat far more than is good for us. Little and often may work satisfactorily so long as it does not develop into much and often, which, needless to say, it is very likely to do. Most people on this account would probably be much better in their health if they ate but twice daily, at noon, and five or six hours before going to bed. Then there is less chance of over-feeding. If, however, we experimentally determine the quantity of food that our particular system requires in order to be maintained in good health, and can trust our self-command in controlling the indulgence of sense, probably the best method is to eat anyway three times daily, and four, five, or even six times, or doing away with set meals altogether, would be a procedure which, [Pg 98] judging from analogy of the anthropoids, ought to be a better method than eating a whole day's supply at once, or at two or three meals.

It is not wise to sit down to a meal when the body is thoroughly fatigued. A glass of hot or cold water will be found reviving, and then, after a short rest, the system will be far better able to assimilate food. When the body is 'tired out,' it stands to reason it cannot perform digestion as easily and as well as when in fit condition.

Also it is unwise to eat immediately before undertaking vigorous muscular work. Strenuous exercise after meals is often the cause of digestive disorders. Starting on exercise after a hearty meal may suspend the gastric digestion, and so prevent the assimilation of protein as to produce a sensation of exhaustion. If, however, rest is taken, the digestive organs proceed with their work, and after a short time recuperation follows, and the exercise can be continued. It is unwise to allow such a suspension of digestion because of the danger of setting up fermentation, or putrefaction, in the food mass awaiting digestion, for this may result in various disorders.

For the same reason it is a bad plan to eat late at night. It is unwise to take a meal just before going to bed, for the digestive organs cannot do their work properly, if at all, while the body is asleep, and the food not being digested is liable to ferment and result in dyspepsia. The 'sinking feeling' sometimes complained of if a [Pg 99] meal is not eaten late at night and described as a kind of hunger is probably due to an abnormal secretion of acid in the stomach. A glass of hot water will often relieve this discomfort. This feeling is seldom experienced by vegetarians of long standing. The natives of India, it is said, do not experience it at all, which fact leads us to surmise the cause to be in some way connected with flesh-eating. Farinaceous foods, however, prepared as soup, porridge, gruel, pultaceous puddings, etc., when eaten, as is customary, without proper insalivation, are liable to be improperly digested and to ferment, giving rise to the sensation described as a 'sinking feeling' and erroneously thought to be hunger.

It is an excellent rule that prescribes fasting when without hunger. When there is no appetite do not eat. It is an example of conventional stupidity that we eat because it is 'meal time,' even though there be not the slightest feeling of genuine hunger. Leaving out of consideration the necessitous poor and those who for their living engage themselves in hard physical toil, it is safe to say that hardly one person in a thousand has ever felt real hunger. Yet no one was ever the worse for waiting upon appetite. No one was ever starved by not eating because of having no appetite. Loss of appetite is a sign that the digestive organs require a rest. It is better to go without food for a time than to force oneself to eat against inclination. The forcing of oneself [Pg 100] to eat to 'keep up one's strength,' is perhaps the quickest way to bring down one's strength by overworking the system and burdening it with material it does not need. Eat by appetite, not by time. Eat frequently when the appetite demands frequent satisfaction, and seldom when seldom hungry. These rules hold good at all times and for everyone. Loss of appetite during sickness should not be looked upon as anything serious in itself, but as a sign that the system does not require food. A sick man like a well man will feel hunger as soon as food is needed, and the practice of tempting the appetite with rich and costly foods is not only a waste of money but is injurious physiologically. Possibly there may be pathological conditions under which hunger cannot make itself felt, but it would seem contrary to Nature as far as the writer, a layman, understands the matter. At least, leaving abnormal conditions of health out of consideration, we can say this much affirmatively: if a man is hungry enough to relish dry bread, then, and then only, does he really require nourishment.

Hunger is always experienced when nutriment is needed, and will be felt a dozen times a day if the food eaten at each of a dozen meals has supplied only sufficient nutriment to produce the force expended between each meal. If the meal is large and supplies sufficient nutriment to produce the force expended in a whole day, then the one meal is all that is required. Never eat to be so [Pg 101] ciable, or conventional, or sensual; eat when hungry.

Professor Pavlov says: 'Appetite is juice'; that is to say, the physiological condition existing when the body has run short of food-fuel, produces a psychological effect, the mind thinking of food, thereby causing through reaction a profuse secretion of saliva, and we say 'the mouth waters.' It is true the appetite is amenable to suggestion. Thus, though feeling hunger, the smell of, or even thought of, decayed food may completely take away appetite and all inclination to eat. This phenomenon is a provision of Nature to protect us from eating impure food. The appetite having thus been taken away will soon return again when the cause of its loss has been removed. Therefore the appetite should be an infallible guide when to eat.

There is one further point to be noted. Food should not be eaten when under the influence of strong emotion. It is true that under such conditions there probably would be no appetite, but when we are so accustomed to consulting the clock that there is danger of cozening ourselves into the belief that we have an appetite when we have not, and so force ourselves to eat when it may be unwise to do so. Strong emotions, as anger, fear, worry, grief, judging by analogy, doubtless inhibit digestive activity. W. B. Cannon, M.D., speaking of experiments on cats, says: 'The stomach movements are inhibited whenever the cat [Pg 102] shows signs of anxiety, rage, or distress.' To thoroughly enjoy one's food, it is necessary to have hunger for it, and if we only eat when we feel hungry, there is little likelihood of ever suffering from dyspepsia.

In passing, it is appropriate to point out that as when food is better enjoyed it is better digested, therefore art, environment, mental disposition, indirectly affect the digestive processes. We should, therefore, remembering that simplicity, not complexity, is the essence of beauty, ornament our food and table, and be as cheerful, sociable, and even as merry as possible. [Pg 103]

IV

HOW TO EAT

The importance of thorough mastication and insalivation cannot be overestimated. The mouth is a part of the digestive apparatus, and in it food is not only broken down, but is chemically changed by the action of the saliva. If buccal (mouth) digestion be neglected, the consequence is that the food passes into the stomach in a condition that renders it difficult for that organ to digest it and any of a great number of disturbances may result.

Mastication means a thorough breaking up of the food into the smallest particles, and insalivation means the mixing of the small particles with the saliva. The mechanical work is done with the jaws and tongue, and the chemical work is performed by the saliva. When the mechanical work is done thoroughly the chemical work is also thorough, and the test for thoroughness is loss of taste. Masticate the food until all taste has disappeared, and then it will be found that the swallowing reflex unconsciously absorbs the food, conscious swallowing, or at least, an effort to swallow, not being called for. [Pg 104]

It may take some while to get into the habit of thorough mastication after having been accustomed to bolting food, but with a conscious effort at the first, the habit is formed, and then the effort is no longer a laborious exercise, but becomes perfectly natural and is performed unconsciously.

This ought to be common knowledge. That such a subject is not considered a necessary part of education is indeed lamentable, for the crass ignorance that everywhere abounds upon the subject of nutrition and diet is largely the cause of the frightful disease and debility so widespread throughout the land, and, as a secondary evil of an enormous waste of labour in the production and distribution of unneeded food. Were everyone to live according to Nature, hygienically and modestly, health, and all the happiness that comes with it, would become a national asset, and as a result of the decreased consumption of food, more time would be available for education, and the pursuit of all those arts which make for the enlightenment and progress of humanity.

To become a convert to this new order, adopting non-animal food and hygienic living, is not synonymous with monastical asceticism, as some imagine. Meat eaters when first confronted with vegetarianism often imagine their dietary is going to be restricted to a monotonous round of carrots, turnips, cabbages, and the like; and if their ignorance prevents them from arguing that it is impossible to maintain health and strength on such [Pg 105] foods, then it is very often objected that carrots and cabbages are not liked, or would not be cared for *all* the time. The best way to answer this objection is to cite a few plain facts. From a catalogue of a firm supplying vegetarian specialties, (and there are now quite a number of such firms), most of the following information is derived:

Of nuts there are twelve varieties, sold either shelled, ground, or in shell. Many of these nuts are also mechanically prepared, and in some cases combined, and made into butters, nut-meats, lard, suet, oil, etc. The varieties of nut-butters are many, and the various combinations of nuts and vegetables making potted savouries, add to a long list of highly nutritious and palatable nut-foods. There are the pulses dried and entire, or ground into flour, such as pea-, bean-, and lentil-flour. There are the cereals, barley, corn, oats, rice, rye, wheat, etc., from which the number of preparations made such as breakfast foods, bread, biscuits, cakes, pastries, etc., is legion. (One firm advertises twenty-three varieties of prepared breakfast foods made from cereals.) Then there are the fruits, fresh, canned, and preserved, about twenty-five varieties; green vegetables, fresh and canned, about twenty-one varieties; and roots, about eleven varieties.

The difficulty is not that there is insufficient variety, but that the variety is so large that there is danger of being tempted beyond the limits dictated by the needs of the body. When, having had [Pg 106] sufficient to eat, there yet remain many highly palatable dishes untasted, one is sometimes apt to gratify sense at the expense of health and good-breeding, to say nothing of economy. Simplicity and purity in food are essential to physical health as simplicity and purity in art are essential to moral and intellectual progress. 'I may say,' says Dr. Haig, 'that simple food of not more than two or three kinds at one meal is another secret of health; and if this seems harsh to those whose day is at present divided between anticipating their food and eating, I must ask them to consider whether such a life is not the acme of selfish shortsightedness. In case they should ever be at a loss what to do with the time and money thus saved from feasting, I would point on the one hand to the mass of unrelieved ignorance, sorrow, and suffering, and on the other to the doors of literature and art, which stand open to those fortunate enough to have time to enter them; and from none of these need any turn aside for want of new Kingdoms to conquer.'

This question of feeding may, by superficial thinkers, be looked upon as unimportant; yet it should not be forgotten that diet has much more to do with health than is commonly realized, and health is intimately connected with mental attitude, and oftentimes is at the foundation of religious and moral development. 'Hypochondriacal crotchets' are often the product of dyspepsia, and valetudinarianism and pessimism are not unrarely [Pg 107] found together. 'Alas,' says Carlyle, 'what is the loftiest flight of genius, the finest frenzy that ever for moments united Heaven with Earth, to the perennial never-failing joys of a digestive apparatus thoroughly eupeptic?'

Our first duty is to learn to keep our body healthy. Naturally, we sooner expect to see a noble character possess a beautiful form than one disfigured by abuse and polluted by disease. We do not say that every sick man is a villain, but we do say that men and women of high character regard the body as an instrument for some high purpose, and believe that it should be cared for and nourished according to its natural requirements. In vegetarianism, *scientifically practised*, is a cure, and better, a preventative, for many physical, mental, and moral obliquities that trouble mankind, and if only a knowledge of this fact were to grow and distil itself into the public mind and conscience, there would be halcyon days in store for future generations, and much that now envelops man in darkness and in sorrow, would be regarded as a nightmare of the past. [Pg 108]

FOOD TABLE

The following table exhibits the percentage chemical composition of the principal vegetable food materials; also of dairy produce and common flesh-foods for comparison.

Food Material	Protein	Fat	Carbohydrates	Salts	Water	Fuel Value cals.
Vegetable Foods	p. ct.	p. ct.	p. ct.	p. ct.	p. ct.	p. lb.
Wheat Flour (entire)	18.8	1.9	71.9	1.0	11.4	1,675
Oatmeal	16.1	7.2	67.5	1.9	7.3	1,860
Rice	8.0	.3	79.0	.4	12.3	1,630
Barley	8.5	1.1	77.8	1.1	11.5	1,650
Corn Meal	9.2	1.9	75.4	1.0	12.5	1,655
Rye	0.8	.9	78.7	.7	12.9	1,630
Lentils (dried)	25.7	1.0	59.2	5.7	8.4	1,620
Beans (dried)	22.5	1.8	59.6	3.5	12.6	1,605
Peas (dried)	24.6	1.0	62.0	2.9	9.5	1,655

Nuts, various (*aver.*)	16.0	52.0	20.0	2.0	10.0	2,640
Dates	2.1	2.8	78.4	1.3	15.4	1,615
Figs	4.3	.3	74.2	2.4	18.8	1,475
Potatoes	2.2	.1	18.4	1.0	78.3	385
Apples	.4	.5	14.2	.3	84.6	290
Bananas	1.3	.6	22.0	.8	75.3	460
Dairy Foods						
Milk, whole (not skim)	3.3	4.0	5.0	.7	87.0	325
Cheese, various (*aver.*)	24.5	28.4	2.1	4.0	41.0	1,779
Hens' Eggs (*boiled*)	14.0	12.0	0.0	.8	73.2	765
Flesh Foods						
Beef	18.6	19.1	0.0	1.0	61.3	1,155
Mutton (*medium fat*)	18.2	18.0	0.0	1.0	62.8	1,105
Ham (*fresh*)	15.6	33.4	0.0	.9	50.1	1,700
Fowl	19.0	16.3	0.0	1.0	63.7	1,045
White Fish (*as purchased*)	22.1	6.5	0.0	1.6	69.8	700

[The amount of heat that will raise one kilogram of water 1 deg. C. is termed a *calorie*. Fuel value, or food units, means the number of calories of heat equivalent to the energy it is assumed the body obtains from food when the nutrients thereof are completely digested.]
[Pg 109]

[Pg 110]

[Pg 111]

ONE HUNDRED RECIPES

RECIPES

The following recipes are given as they appear in the English edition of this book and were prepared for English readers. While some of these will be difficult for American readers to follow, we give them as in the original edition, and many of the unusual ingredients called for can be obtained from the large grocers and dealers, and if not in stock will be obtained to order. 'Nutter' is a name given a nut butter used for cooking. It is, so far as we know, the only collection of strictly vegetarian recipes published.

Readers interested in the foreign products referred to, should write to Pitman's Health Food Company, Aston Brook St., Birmingham, England, and to Mapleton's Nut Food Company, Ltd., Garston, Liverpool, England, for price list and literature.

The Publishers.
[Pg 112]

SOUPS

1. — Vegetable Soup

1 large cupful red lentils, 1 turnip, 2 medium onions, 3 potatoes, 1 carrot, 1 leek, 1 small head celery, parsley, 1 lb. tomatoes, 3½ quarts water.

Wash and cut up vegetables, but do not peel. Boil until tender, then strain through coarse sieve and serve. This soup will keep for several days and can be reheated when required.

2. — Semolina Soup

4 oz. semolina, 2 chopped onions, 1 tablespoonful gravy essence, [6] 2 quarts water or vegetable stock. [7]

3. – Spinach Soup No. 1

1 lb. Spinach, 1 tablespoonful gravy essence, 1 quart water.

Cook spinach in its own juices (preferably in double boiler). Strain from it, through a hair sieve or colander, all the liquid. Add essence and serve.

4. – Spinach Soup No. 2

1 lb. spinach, 1 lb. can tomatoes, 1 tablespoonful nut-milk (Mapleton's), 1½ pints water.

Dissolve nut-milk in little water, cook all ingredients together in double-boiler for 1½ hours, strain and serve.

5. – Pea Soup

4 ozs. pea-flour, 2 potatoes, 1 large onion, 1 tablespoonful gravy essence, 2 quarts water.

Cook potatoes, (not peeled), and onion until soft. Skin and mash potatoes and chop onion. Mix pea-flour into paste with little water. Boil all ingredients together for 20 minutes, then serve.

Lentil and Haricot Soups

These are prepared in the same way as Recipe No. 5 substituting lentil, or haricot flour for pea-flour. [Pg 113]

6. – Tomato-Pea Soup

4 ozs. pea-flour, 1 lb. tin tomatoes, 1 chopped leek, 1 quart water.

Mix pea-flour into paste with little water. Boil ingredients together 30 minutes, then serve.

Tomato-Lentil and Tomato-Bean Soups

These are prepared in the same way as Recipe No. 6, substituting lentil-, or bean-flour for pea-flour.

7. — Rice-Vermicelli Soup

2 ozs. rice-vermicelli, 1 tablespoonful nut-milk, 1 dessertspoonful gravy essence, 1 quart water.

Boil vermicelli in water until soft. Dissolve nut-milk in little water. Boil all ingredients together 5 minutes, then serve.

8. — Pea-Vermicelli Soup

2 ozs. pea-vermicelli, 1 tablespoonful nut-milk, 1 tablespoonful tomato purée, 1 quart water.

Boil vermicelli in water until soft, dissolve nut-milk in little water. Boil all ingredients together 5 minutes, then serve.

9. — Pot-barley Soup No. 1

4 ozs. pot-barley, 1 onion, 1 tablespoonful gravy essence, 2 quarts water, corn flour to thicken.

Cook barley until quite soft; chop onion finely; mix a little corn flour into paste with cold water. Stir into the boiling soup. Boil all ingredients together for 20 minutes, then serve.

Wheat and Rice Soups

These are prepared in the same way as Recipe No. 9, substituting wheat or rice grains for barley.

10. — Pot-barley Soup No. 2

4 ozs. pot-barley, 1 dessertspoonful nut-milk, 1 chopped onion, 1 dessertspoonful tomato purée, 1 quart water.

Cook barley until soft; dissolve nut-milk in little water; boil all ingredients together for 20 minutes, then serve.

11. — Corn Soup

1 lb. tin sugar-corn, ½ lb. tin tomatoes, 2 chopped onions, 2 ozs. corn flour, 1 quart water.

Boil onion until soft; mix corn flour into paste with cold water. Place sugar-corn, tomatoes, onions, and water into stew pan; heat and add corn flour. Boil ingredients together 10 minutes, and serve. [Pg 114]

SAVORY DISHES

12. — Nut Rissoles

3 ozs. mixed grated nuts, 3 ozs. breadcrumbs, 1 oz. nut butter, 1 chopped onion, 1 large cupful canned tomatoes.

Mix ingredients together; mould into rissoles, dust with flour and fry in 'Nutter.' Serve with gravy.

13. — Lentil Cakes

8 ozs. red lentils, 3 ozs. 'Grape Nuts,' 1 small onion, 1 teaspoonful gravy essence, breadcrumbs.

Cook lentils until soft in smallest quantity of water; chop onion finely; mix all ingredients, using sufficient breadcrumbs to make into stiff paste; form into cakes and fry in 'Nutter.' Serve with gravy.

14. — Marrow Roast

1 vegetable marrow, 3 ozs. grated nuts, 1 onion, 1 oz. 'Nutter,' 1 cup breadcrumbs, 2 teaspoonfuls tomato purée.

Cook marrow, taking care not to allow it to break; when cold, peel, cut off one end and remove seeds with spoon. Prepare stuffing:—chop onion finely; melt nut fat and mix ingredients together. Then stuff marrow and tie on decapitated end with tape; sprinkle with breadcrumbs and bake 30 minutes. Serve with gravy.

15.—Stewed Celery

1 head celery, 4 slices whole-meal bread, nut butter.

Slice celery into suitable lengths, which steam until soft. Toast and butter bread, place celery on toast and cover with pea, bean, or lentil sauce, (see Recipe No. 39).

16.—Barley Entrée

4 ozs. pot-barley, 1 lb. tin tomatoes, 1 chopped onion, 2 tablespoonfuls olive oil.

Cook barley until quite soft in smallest quantity of water (in double boiler). Then add tomatoes and oil, and cook for 10 minutes. To make drier, cook barley in tomato juice adding only 2 or 3 tablespoonfuls of water.

Rice, Wheat, Macaroni, Lentil, Bean, Split-pea Entrées

These are prepared in the same way as Recipe No. 16, substituting one of these cereals or légumes for barley.

17.—Savory Pie

Paste (Recipe No. 59), marrow stuffing (Recipe No. 14).

Line sandwich tin with paste; fill interior with stuffing; cover with paste or cooked sliced potatoes; bake in sharp oven. [Pg 115]

18.—Baked Bananas

Prepare the desired number by washing and cutting off stalk, but do not peel. Bake in oven 20 minutes, then serve.

19. — Barley Stew

4 ozs. pot-barley, 2 onions, parsley.

Chop onions and parsley finely; cook ingredients together in very small quantity of water in double boiler until quite soft. Serve with hot beetroot, or fried tomatoes or potatoes.

Corn, Rice, Frumenty, Pea-Vermicelli Stews

These are prepared in the same way as Recipe No. 19, substituting one of the above cereals or pulses for barley.

20. — Mexican Stew

1 cupful brown beans, 2 onions, 2 potatoes, 4 tomatoes, 1 oz. sugar, 1 cupful red grape-juice, rind of 1 lemon, water.

Soak beans overnight; chop vegetables in chunks; boil all ingredients together 1 hour.

21. — Vegetable Pie

5 ozs. tapioca, 4 potatoes, 3 small onions, paste, (see Recipe No. 59), tomato purée to flavor.

Soak tapioca. Partly cook potatoes and onions, which then slice. Place potatoes, onions, and tapioca in layers in pie-dish; mix purée with a little hot water, which pour into dish; cover with paste and bake.

22. — Rice Rissoles

6 ozs. unpolished rice, 1 chopped onion, 1 dessertspoonful tomato purée, breadcrumbs.

Boil rice and onion until soft; add purée and sufficient breadcrumbs to make stiff; mould into rissoles; fry in 'Nutter,' and serve with parsley sauce, (Recipe No. 38).

23. — Scotch Stew

3 ozs. pot-barley, 2 ozs. rolled oats, 1 carrot, 1 turnip, 2 potatoes, 1 onion, 4 tomatoes, water.

Wash, peel, and chop vegetables in chunks. Stew all ingredients together for 2 hours. Dress with squares of toasted bread.

24. — Plain Roasted Rice

Steam some unpolished rice until soft; then distribute thinly on flat tin and brown in hot oven.

25. — Nut Roast No. 1

1 lb. pine kernels (flaked), 4 tablespoonfuls pure olive oil, 2 breakfastcupfuls breadcrumbs, ½ lb. tomatoes (peeled and mashed). [Pg 116]

Mix ingredients together, place in pie-dish, sprinkle with breadcrumbs, and bake until well browned.

26. — Nut Roast No. 2

1 lb. pine kernels (flaked), 1 cooked onion (chopped), ½ cupful chopped parsley, 8 ozs. cooked potatoes (mashed).

Mix ingredients together, place in pie-dish and cover with layer of boiled rice. Cook until well browned.

27. — Maize Roast

8 ozs. corn meal, 1 large Spanish onion (chopped), 2 tablespoonfuls nut-milk, 1 dessertspoonful gravy essence.

Cook onion; dissolve nut-milk thoroughly in about ½ pint water.

Boil onion, nut-milk, and essence together two minutes, then mix all ingredients together, adding sufficient water to make into very soft batter; bake 40 minutes.

28. — Plain Savory Rice

4 ozs. unpolished rice, 1 lb. tin tomatoes.

Boil together until rice is cooked. If double boiler be used no water need be added, and thus the rice will be dry and not pultaceous.

29. — Potato Balls

4 medium sized potatoes, 1 large onion (chopped), 1 dessertspoonful pure olive oil, breadcrumbs.

Cook onion and potatoes, then mash. Mix ingredients, using a few breadcrumbs and making it into a very soft paste. Roll into balls and fry in 'Nutter,' or nut butter.

30. — Bean Balls

4 ozs. brown haricot flour, 1 onion (chopped), 1 dessertspoonful pure olive oil, 1 tablespoonful tomato purée, breadcrumbs.

Cook onion; mix flour into paste with purée and oil; add onion and few breadcrumbs making into soft paste. Fry in 'Nutter.'

31. — Lentil and Pea Balls

These are made in the same way as Recipe No. 30, substituting lentil-or pea-flour for bean-flour.

31. — Lentil Patties

4 ozs. lentils, 1 small onion (chopped), 1 oz. 'Nutter,' or nut butter, 1 teaspoonful gravy essence, paste (see Recipe No. 59).

Cook ingredients for filling all together until lentils are quite soft. Line patty pans with paste; fill, cover with paste and bake in sharp oven.

Barley, Bean, Corn, Rice, and Wheat Patties

These are prepared in the same way as in Recipe No. 31, [Pg 117] substituting one of the above cereals or beans for lentils.

32. — Lentil Paste

8 ozs. red lentils, 1 onion (chopped), 4 tablespoonfuls pure olive oil, breadcrumbs.

Boil lentils and onions until quite soft; add oil and sufficient breadcrumbs to make into paste; place in jars; when cool cover with melted nut butter; serve when set.

33. — Bean Paste

8 ozs. small brown haricots, 2 tablespoonfuls tomato purée, 1 teaspoonful 'Vegeton,' 2 ozs. 'Nutter' or nut butter, 1 cup breadcrumbs.

Soak beans over night; flake in Dana Food Flaker; place back in fresh water and add other ingredients; cook one hour; add breadcrumbs, making into paste; place in jars, when cool cover with nut butter; serve when set.

34. — Spinach on Toast

Cook 1 lb. spinach in its own juice in double boiler. Toast and butter large round of bread. Spread spinach on toast and serve. Other vegetables may be served in the same manner.

GRAVIES AND SAUCES

35. — Clear Gravy

1 teaspoonful 'Marmite,' 'Carnos,' 'Vegeton,' or 'Pitman's Vigar Gravy Essence,' dissolved in ½ pint hot water.

36. — Tomato Gravy

1 teaspoonful gravy essence, 1 small tablespoonful tomato purée, ½ pint water. Thicken with flour if desired.

37. — Spinach Gravy

1 lb. spinach, 1 dessertspoonful nut-milk, ½ pint water.

Boil spinach in its own juices in double boiler; strain all liquid from spinach and add it to the nut-milk which has been dissolved in the water.

38. — Parsley Sauce

1 oz. chopped parsley, 1 tablespoonful olive oil, a little flour to thicken, ½ pint water.

39. — Pea, Bean, and Lentil Sauces

1 teaspoonful pea-, or bean-, or lentil-flour; ½ teaspoonful gravy essence, ½ pint water.

Mix flour into paste with water, dissolve essence, and bring to a boil. [Pg 118]

PUDDINGS, ETC.

40. — Fig Pudding

1 lb. whole-meal flour, 6 ozs. sugar, 6 ozs. 'Nutter,' or nut butter, ½ chopped figs, 1 teaspoonful baking powder, water.

Melt 'Nutter,' mix ingredients together with water into stiff batter; place in greased pudding basin and steam 2 hours.

31. — Date Pudding

1 lb. breadcrumbs, 6 ozs. sugar, 6 ozs. 'Nutter,' ½ lb. stoned and chopped dates, 1 teaspoonful baking powder, water.

Melt 'Nutter'; mix ingredients together with water into stiff batter; place in greased pudding basin and steam 2 hours.

Prune, Ginger, and Cherry Puddings

These are prepared the same way as in Recipe No. 40, or No. 41, substituting prunes or preserved ginger, or cherries for figs or dates.

42. – Rich Fruit Pudding

1 lb. whole-meal flour, 6 ozs. almond cream, 6 ozs. sugar, 3 ozs. preserved cherries, 3 ozs. stoned raisins, 3 ozs. chopped citron, 1 teaspoonful baking powder, water.

Mix ingredients together with water into stiff batter; place in greased pudding basin and steam 2 hours.

43. – Fruit-nut Pudding No. 1

½ lb. white flour, ¼ lb. whole meal flour, ¼ lb. mixed grated nuts, 6 ozs. 'Nutter' or nut butter, 6 ozs. sugar, 6 ozs. sultanas, 2 ozs. mixed peel (chopped), 1 teaspoonful baking powder, water.

Melt nut-fat, mix ingredients together with water into stiff batter; place in greased pudding basin and steam 2 hours.

44. – Fruit-nut Pudding No. 2

½ lb. white flour, ¼ lb. ground rice, ¼ lb. corn meal, 4 ozs. chopped dates or figs, 4 ozs. chopped almonds, 6 ozs. almond nut-butter, 6 ozs. sugar, 1 teaspoonful baking powder, water.

Melt butter, mix ingredients together with water into stiff batter; place in greased pudding basin and steam 2 hours.

45. – Maize Pudding No. 1

½ lb. maize meal, 3 ozs. white flour, 3 ozs. 'Nutter,' 3 ozs. sugar, ½ tin pineapple chunks, 1 teaspoonful baking powder. [Pg 119]

Melt fat, cut chunks into quarters; mix ingredients with very little water into batter; place in greased pudding basin and steam 2 hours.

46. — Maize Pudding No. 2

6 ozs. corn meal, 3 ozs. white flour, 2 ozs. 'Nutter,' 2 ozs. sugar, 3 tablespoonfuls marmalade, 1 teaspoonful baking powder, water.

Melt 'Nutter,' mix ingredients together with little water into batter; place in greased pudding basin and steam 2 hours.

47. — Cocoanut Pudding

6 ozs. whole wheat flour, 2 ozs. cocoanut meat, 2 ozs. 'Nutter,' 2 ozs. sugar, 1 small teaspoonful baking powder, water.

Melt fat, mix ingredients together with water into batter; place in greased pudding basin and steam 2 hours.

48. — Tapioca Apple

1 cup tapioca, 6 large apples, sugar to taste, water.

Soak tapioca, peel and slice apples; mix ingredients together, place in pie-dish with sufficient water to cover and bake.

49. — Oatmeal Moulds

4 ozs. rolled oats, 2 ozs. sugar, 4 ozs. sultanas, water.

Cook oatmeal thoroughly in double boiler, then mix ingredients together; place in small cups, when cold turn out and serve with apple sauce, or stewed prunes.

50. — Carrot Pudding

4 ozs. breadcrumbs, 4 ozs. 'Nutter,' 4 ozs. flour, 4 ozs. mashed carrots, 4 ozs. mashed potatoes, 6 ozs. chopped raisins, 2 ozs. brown sugar, 1 dessertspoonful treacle, 1 teaspoonful baking powder.

Mix ingredients well, place in greased pudding basin and steam 2 hours.

51. — Sultana Pudding

½ lb. whole meal flour, 1 breakfastcupful breadcrumbs, 4 ozs. ground pine kernels, pignolias or almonds, ½ lb. sultanas, 4 ozs. sugar, water.

Mix ingredients together into a stiff batter; place in greased basin and steam 2 hours.

52. — Semolina Pudding

4 ozs. semolina, 1 oz. corn flour, 3 ozs. sugar, rind of one lemon, 1½ pints water.

Mix corn flour into paste in little water; place ingredients in double boiler and cook for 1 hour, place in pie-dish and brown in sharp oven. [Pg 120]

53. — Rice Mould

4 ozs. ground rice, 1 oz. sugar, ½ pint grape-juice.

Cook ingredients in double boiler, place in mould. When cold turn out and serve with stewed fruit.

54. — Maize Mould

6 ozs. corn meal, 2 ozs. sugar, ½ pint grape-juice, 1½ pints water.

Cook ingredients in double boiler for 1 hour; place in mould. When cold turn out and serve with stewed fruit.

55. — Lemon Sago

4 ozs. sago, 7 ozs. golden syrup, juice and rind of two lemons, 1½ pints water.

Boil sago in water until cooked, then mix in other ingredients. Place in mould, turn out when cold.

56. — Lemon Pudding

4 ozs. breadcrumbs, 1 oz. corn flour, 2 ozs. sugar, rind one lemon, 1 pint water.

Mix corn flour into paste in little water; mix ingredients together, place in pie-dish, bake in moderate oven.

57. — Prune Mould

1 lb. prunes, 4 ozs. sugar, juice 1 lemon, ¼ oz. agar-agar, 1 quart water.

Soak prunes for 12 hours in water, and then remove stones. Dissolve the agar-agar in the water, gently warming. Boil all ingredients together for 30 minutes, place in mould, when cold turn out and decorate with blanched almonds.

58. — Lemon Jelly

¼ oz. agar-agar, 3 ozs. sugar, juice 3 lemons, 1 quart water.

Soak agar-agar in the water for 30 minutes; add fruit-juice and sugar, and heat gently until agar-agar is completely dissolved, pour into moulds, turn out when cold.

This jelly can be flavoured with various fruit juices, (fresh and canned). When the fruit itself is incorporated, it should be cut up into small pieces and stirred in when the jelly commences to thicken. The more fruit juice added, the less water must be used. Such fruits as fresh strawberries, oranges, raspberries, and canned pineapples, peaches, apricots, etc., may be used this way.

59. — Pastry

1 lb. flour, ½ lb. nut-butter or nut fat, 2 teaspoonfuls baking powder, water.

Mix with water into stiff paste. This is suitable for tarts, patties, pie-covers, etc. [Pg 121]

CAKES

60. — Wheatmeal Fruit Cake

6 ozs. entire wheat flour, 3 ozs. nut-butter, 3 ozs. sugar, 3 ozs. almond meal, 10 ozs. sultanas, 2 ozs. lemon peel, 2 teaspoonsful baking powder.

Rub butter into flour, mix all ingredients together with water into stiff batter; bake in cake tins lined with buttered paper.

61. — Rice Fruit Cake

8 ozs. ground rice, 4 ozs. white flour, 4 ozs. 'Nutter,' 3 ozs. sugar, 6 ozs. stoned, chopped raisins, 1 large teaspoonful baking powder, water.

Rub 'Nutter' into flour, mix all ingredients together with water into stiff batter; bake in cake tins lined with buttered paper.

62. — Maize Fruit Cake

8 ozs. corn meal, 6 ozs. white flour, 4 ozs. sugar, 4 ozs. nut-butter, 8 ozs. preserved cherries, 2 ozs. lemon peel, 2 teaspoonfuls baking powder, water.

Rub butter into flour, mix all ingredients together with water into stiff batter; bake in cake tins lined with buttered paper.

63. — Apple Cake

1 lb. apples, ¼ lb. white flour, ½ lb. corn meal, 4 ozs. 'Nutter,' 4 ozs. sugar, 2 small teaspoonfuls baking powder, water.

Cook apples to a sauce and strain well through colander, rejecting lumps. Melt fat and mix all ingredients together with water into stiff batter; bake in cake tins lined with buttered paper.

64. — Corn Cake (plain)

½ lb. maize meal, 3 ozs. 'Nutter,' 3 ozs. sugar, 1 teaspoonful baking powder.

Melt fat, mix all ingredients together into batter; bake in cake tins lined with buttered paper.

65. — Nut Cake

12 ozs. white flour, 4 ozs. ground rice, 4 ozs. 'Nutter,' or nut butter, 5 ozs. sugar, 6 ozs. mixed grated nuts, 2 teaspoonfuls baking powder.

Melt fat, mix ingredients together into batter, and place in cake tins lined with buttered paper. [Pg 122]

66. — Mixed Fruit Salads

2 sliced bananas, 1 tin pineapple chunks, 2 sliced apples, 2 sliced oranges, ½ lb. grapes, ¼ lb. raisins, ¼ lb. shelled walnuts, ½ pint grape-juice.

67. — Fruit Nut Salad

1 lb. picked strawberries, ¼ lb. mixed shelled nuts, ½ pint grape-juice. Sprinkle over with 'Granose' or 'Toasted Corn Flakes' just before serving.

68. — Winter Salad

2 peeled, sliced tomatoes, 2 peeled, sliced apples, 1 small sliced beetroot, 1 small sliced onion, olive oil whisked up with lemon juice for a dressing.

69. — Vegetable Salad

1 sliced beetroot, 1 sliced potato (cooked), 1 sliced onion, 1 sliced heart of cabbage, olive oil dressing; arrange on a bed of water-cress.

BISCUITS

The following biscuits are made thus:—Melt the 'Nutter,' mix all ingredients with sufficient water to make into stiff paste; roll out and cut into shapes. Bake in moderate oven.

These biscuits when cooked average 20 grains protein per ounce.

70. — Plain Wheat Biscuits

½ lb. entire wheat flour, 4 ozs. sugar, 4 ozs. 'Nutter,' little chopped peel.

71. — Plain Rice Biscuits

3-4 lb. ground rice, 4 ozs. sugar, 3 ozs. 'Nutter,' vanilla essence.

72. — Plain Maize Biscuits

½ lb. maize meal, 4 ozs. sugar, 3 ozs. 'Nutter.'

(If made into soft batter these can be dropped like rock cakes).

73. — Banana Biscuits

½ lb. banana meal, 4 ozs. sugar, 4 ozs. 'Nutter.' [Pg 123]

74. — Cocoanut Biscuits

½ lb. white flour, 3 ozs. sugar, 2 ozs. 'Nutter,' 4 ozs. cocoanut meal.

75. — Sultana Biscuits

3-4 lb. white flour, 4 ozs. sugar, 4 ozs. 'Nutter,' 6 ozs. minced sultanas and peel 2 ozs. almond meal.

78. — Fig Biscuits

½ lb. entire wheat flour, 3 ozs. sugar, 4 ozs. 'Nutter,' 3 ozs. minced figs.

(If made into soft batter these can be dropped like rock cakes).

Date, Prune, Raisin, and Ginger Biscuits

These are prepared in the same way as Recipe No. 76, using one of these fruits in place of figs. (Use dry preserved ginger).

77. — Brazil-nut Biscuits

8 ozs. white flour, 2 ozs. ground rice, 3 ozs. sugar, 4 ozs. grated brazil kernels.

(If made into a soft batter these can be dropped like rock cakes).

78. — Fruit-nut Biscuits

¾ lb. white flour, 4 ozs. ground rice, 4 ozs. sugar, 5 ozs. 'Nutter,' 6 ozs. mixed grated nuts, 6 ozs. mixed minced fruits, sultanas, peel, raisins.

79. — Rye Biscuits

1 lb. rye flour, 8 ozs. sugar, 8 ozs. nut butter, 8 ozs. sultanas.

80. — Xerxes Biscuits

¾ lb. whole wheat flour, 2 ozs. sugar, ½ breakfastcupful olive oil.

BREADS (unleavened)

These are prepared as follows: Mix ingredients with water into stiff dough; knead well, mould, place in bread tins, and bake in slack oven for from 1½ to 2½ hours (or weigh off dough into ½ lb. pieces, mould into flat loaves, place on flat tin, cut across diagonally with sharp knife and bake about 1½ hours).

81. — Apple Bread

2 lbs. entire wheat meal doughed with 1 lb. apples, cooked in water to a pulp.

82. — Rye Bread

2 lbs. rye flour, ¾ lb. ground rice. [Pg 124]

83. — Plain Wheat Bread

2 lbs. finely ground whole wheat flour.

84. — Corn Wheat Bread

1 lb. whole wheat flour, 1 lb. cornmeal.

85. — Rice Wheat Bread

1 lb. ground rice, 1 lb. whole wheat flour, 1 lb. white flour.

86. — Date Bread

2 lbs. whole wheat flour, ¾ lb. chopped dates.

87. — Ginger Bread

¾ lb. whole wheat flour, ¾ lb. white flour, ¼ lb. chopped preserved ginger, a little cane sugar.

88. — Cocoanut Bread

1 lb. whole wheat flour, 1 lb. white flour, ½ lb. cocoanut meal, some cane sugar.

89. — Fig Bread

1½ lbs. whole wheat flour, ½ lb. white flour, ½ lb. chopped figs.

90. — Sultana Bread

½ lb. ground rice, ½ lb. maize meal, ½ lb. white flour, ½ lb. sultanas.

91. — Fancy Rye Bread

1½ lbs. rye flour, ½ lb. currants and chopped peel, a little cane sugar.

PORRIDGES

92. — Maize, Meal, Rolled Oats, Ground Rice, etc., thoroughly cooked make excellent porridge. Serve with sugar and unfermented fruit-juice.

FRUIT CAKES

The following uncooked fruit foods are prepared thus: Mix all ingredients well together; roll out to ¼ inch, or ½ inch, thick; cut out with biscuit cutter and dust with ground rice.

93. — Date Cakes

1½ lbs. stoned dates minced, ½ lb. mixed grated nuts.

94. — Fig Cakes

1½ lbs. figs minced, ½ lb. ground almonds. [Pg 125]

95. — Raisin-Nut Cakes

½ lb. stoned raisins minced, 6 ozs. mixed grated nuts.

96. — Ginger-Nut Cakes

½ lb. preserved ginger (minced), ½ lb. mixed grated nuts. 4 ozs. 'Grape Nuts.'

97. — Prune-Nut Cakes

½ lb. stoned prunes (minced), ½ lb. grated walnuts.

98. — Banana-Date Cakes

8 ozs. figs (minced); 4 bananas; sufficient 'Wheat or Corn Flakes' to make into stiff paste.

100. — Cherry-Nut Cakes

8 ozs. preserved cherries (minced); ½ lb. mixed grated nuts; sufficient 'Wheat or Corn Flakes' to make into stiff paste. [Pg 126]

FOOTNOTES:

[1] It seems reasonable to suppose that granting the organism has such natural needs satisfied as sleep, warmth, pure air, sunshine, and so forth, fundamentally all susceptibility to disease is due to wrong feeding and mal-nutrition, either of the individual organism or of its progenitors. The rationale of nutrition is a far more complicated matter than medical science appears to realise, and until the intimate relationship existing between nutrition and pathology has been investigated, we shall not see much progress towards the extermination of disease. Medical science by its curative methods is simply pruning the evil, which, meanwhile, is sending its roots deeper into the unstable organisms in which it grows.

[2] See *Sartor Resartus*, Book I., chap. xi.: Book III., chap. vii. Also an article by Prof. W. P. Montague, Ph.D.: 'The Evidence of Design in the Elements and Structure of the Cosmos,' in the *Hibbert Journal*, Jan., 1904.

[3] This is not an exaggeration. 'Genoa Cake,' for instance, contains ten varieties of food: butter, sugar, eggs, flour, milk, sultanas, orange and lemon peel, almonds, and baking powder.

[4] Entire-wheat flour averages .9 per cent. fibre; high-grade white flour, .2 per cent. fibre.

[5] See United States Dept. of Agriculture, Farmer's Bulletin, No. 249, page 19, obtainable from G. P. O., Washington, D. C.

[6] There are several brands of wholly vegetable gravy essence now on the market. The best known are 'Vegeton,' 'Marmite,' 'Carnos,' and Pitman's 'Vigar Gravy Essence.'

[7] Vegetable stock is the water that vegetables have been boiled in; this water contains a certain quantity of valuable vegetable salts, and should never be thrown away.

The Health Culture Co.

For more than a dozen years the business of the Health-Culture Co. was conducted in New York City, moving from place to place as increased room was needed or a new location seemed to be more desirable.

In 1907 the business was removed to Passaic, N. J., where it is pleasantly and permanently located in a building belonging to the proprietor of the company.

There has never been as much interest in the promotion and preservation of personal health as exists to-day. Men and women everywhere are seeking information as to the best means of increasing health and strength with physical and mental vigor.

HEALTH-CULTURE, a monthly publication devoted to Practical Hygiene and Bodily Culture, is unquestionably the best publication of its kind ever issued. It has a large circulation and exerts a wide influence, numbering among its contributors the best and foremost writers on the subject.

THE BOOKS issued and for sale by this Company are practical and include the very best works published relating to Health and Hygiene.

THE HEALTH APPLIANCES, manufactured and for sale, include Dr. Forest's Massage Rollers and Developers, Dr. Wright's Colon Syringes, the Wilhide Exhaler, etc. and we are prepared to furnish anything in this line, Water-Stills, Exercisers, etc.

CIRCULARS and price lists giving full particulars will be sent on application.

INQUIRIES as to what books to read or what appliances to procure for any special conditions cheerfully and fully answered. If you have any doubts state your case and we will tell you what will best meet it. If you want books of any kind we can supply them at publisher's prices.

Address
THE HEALTH-CULTURE CO.,

Turner Building, Passaic, N. J.
[Pg 127]

DR. FOREST'S Massage Rollers

Dr. Forest is the inventor and originator of Massage Rollers, and these are the original and only genuine Massage Rollers made. The making of others that are infringements on our patents have been stopped or they are inferior and practically worthless. In these each wheel turns separately, and around the centre of each is a band or buffer of elastic rubber.

The rollers are made for various purposes, each in a style and size best adapted for its use, and will be sent prepaid on receipt of price.

No. 1. Six Wheels, Body Roller, $2.

The best size for use over the body, and especially for indigestion, constipation, rheumatism, etc. Can also be used for reduction.

No. 2, Four Wheels, Body Roller, $1.50.

Smaller and lighter than No. 1; for small women it is the best in size, for use over the stomach and bowels, the limbs, and for cold feet.

No. 3, Three Wheels, Scalp Roller, $1.50.

Made in fine woods and for use over the scalp, for the preservation of the hair. Can be used also over the neck to fill it out and for the throat.

No. 4, Five Wheels, Bust Developer, $2.50.

The best developer made. By following the plain physiological directions given, most satisfactory results can be obtained.

No. 5, Twelve Wheels, Abdominal Roller, $4.

For the use of men to reduce the size of the abdomen, and over the back. The handles give a chance for a good, firm, steady, pressure.

No. 6, Three Small Wheels, Facial Roller, $2.50.

Made in ebony and ivory, for use over the face and neck, for preventing and removing wrinkles, and restoring its contour and form.

No. 7, Three Wheels, Facial Massage Roller, $1.50.

Like No. 6, made in white maple. In other respects the same.

No. 8, Eight Wheels, Abdominal Roller, $3.50.

This is the same as No. 5, except with the less number of wheels. Is made for the use of women, for reducing hip and abdominal measure.

With each roller is sent Dr. Forest's Manual of Massotherapy; containing 100 pages, giving full directions for use. Price separately 25c.
[Pg 128]

THE ATTAINMENT OF EFFICIENCY

Rational Methods of Developing Health and Personal Power

By W. R. C. Latson M. D., Author of "Common Disorders," "The Enlightened Life," Etc.

This work by Dr. Latson indicates the avenues that lead to efficient and successful living, and should be read by every man and woman who would reach their best and attain to their highest ambitions in business, professional, domestic or social life. Something of the scope of this will be seen from the following

TABLE OF CONTENTS.

How to Live the Efficient Life.—Man a Production of Law—Determining Factors in Health and Power—The Most Wholesome Diet—Practical Exercises for Efficiency—Influence of Thought Habits.

Mental Habits and Health.—All is Mind—Seen in Animals—Formative Desire in the Jungle—Mind the Great Creator—Mind the One Cause of Disease—Faulty Mental Habits.

The Conquest of Worry.—Effects Upon Digestion—Anarchy of the Mind—A Curable Disorder.

Secret of Mental Supremacy.—Practical Methods—The Key Note—Mental Power a Habit.

The Nobler Conquest.—Life a Struggle—Who Are the Survivors?—The Art of Conquest—The Struggle with the World—Effects of Opposition.

Firmness One Secret of Power.—Without Firmness no Real Power—How it Grows with Exercise—Gaining the Habit of Firmness.

Self-Effacement and Personal Power.—Growing Older in Wisdom—The Fallacy of Identity—Self-Preservation the First Law.

The Power of Calmness.—The Nervous System—Effects of Control.

How to Be an Efficient Worker.—How to Work—Making Drudgery a Work of Art.

The Attainment of Personal Power.—An Achievement—Know Yourself—Learning from Others.

The Secret of Personal Magnetism.—What is Personal Magnetism?—Effects of the Lack of It—How to Gain It.

The Prime Secret of Health.—What is Essential?—What to Do—How to Do It.

How to Increase Vitality.—The Mark of the Master—What Is Vitality?—Possibility of Increase—Spending Vitality.

The Attainment of Physical Endurance.—Essential to Success—The Secret of Endurance—Working Easily—Economizing Strength—Exercises for Promoting Endurance.

The Attainment of Success.—The Secret of Success—What to Do to Acquire It.

The Way to Happiness.—A Royal Road to Happiness—The Secret of Happiness.

How to Live Long in the Land.—Characteristics—Essentials—Bodily Peculiarities.

The Gospel of Rest.—All Need It—Few get It—The Secret of Rest—Its Effects.

Sleeping as a Fine Art.—Causes of Sleeplessness—The Mind. How to Control It.

Common Sense Feeding.—What is Proper Feeding?—Many Theories—Mental Conditions—The Kind of Food.

Grace and How to Get It.—What is Grace—Hindrances to Grace—Exercises for Grace.

Style and How to Have It.—The Secret of Style—Carriage of the Body—Exercises for Stylishness.

How to Have a Fine Complexion.—What Effects the Complexion?—The Secret of a Good Complexion—Effects of Food.

The Secret of a Beautiful Voice.—What the Voice Is—Easily Acquired.

How to Cure Yourself When Sick.—It is Easy—What is Disease?—Nature's Efforts—Best Remedies.

One of the most practical and helpful works published on personal improvement and the acquiring of physical and mental vigor; a key to efficient manhood and womanhood and a long, happy and helpful life. All who are striving for success should read it.

Artistically bound in Ornithoid covers. Price 50c. An extra edition is issued on heavy paper, bound in fine cloth. Price $1.00.

[Pg 129]

WOMANLY BEAUTY

In Form and Features.

Containing specially written chapters from well-known authorities on the cultivation of personal beauty in women, as based upon Health-Culture; fully illustrated. Edited by Albert Turner. Bound in extra cloth, price; $1.00.

This is the best and most comprehensive work ever published on Beauty Culture, covering the entire subject by specialists in each department, thus giving the work a greatly increased value. It is profusely and beautifully illustrated; a handsome volume. Some idea of the scope of this may be seen from the

TABLE OF CONTENTS.

Introduction. By Ella Van Poole.

Womanly Beauty: Its Requirements. By Dr. Jacques.

Why It Lasts or Fades. By Dr. C. H. Stratz.

Temperamental Types. By Sarah C. Turner.

Breathing and Beauty. By Dr. W. R. C. Latson.

Curative Breathing. By Madame Donna Madixxa.

Sleep; Its Effect on Beauty. By Ella Van Poole.

The Influence of Thought Upon Beauty. By Dr. W. R. C. Latson.

Health and Beauty. By Dr. Chas. H. Shepard.

The Home A Gymnasium. By Mrs. O. V. Sessions.

Facial Massage. By Ella Van Poole.

The Hair; Its Care and Culture. By Albert Turner.

Care of the Hands and Feet. By Stella Stuart.

Exercising for Grace and Poise. Illustrated.

A Good Form, and How to Secure It. From Health-Culture.

How to Have a Good Complexion. By Susanna W. Dodds M. D.

Bust Development; How to Secure It.

Exercise: Who Needs It; How to Take It. Edward B. Warman.

Perfumes and Health. By Felix L. Oswald, M. D.

The Voice as an Element of Beauty. By Dr. Latson.

How to be Beautiful. By Rachel Swain, M. D.

The Ugly Duckling. A Story. By Elsie Carmichael.

Dress and Beauty. By Ella Van Poole.

Some Secrets About a Beautiful Neck. By Eleanor Wainwright.

Hints in Beauty Culture. Compiled By The Editor.

It is an encyclopedia on the subject, covering every phase of the question in a practical way, and should be in the hands of every woman who would preserve her health and personal appearance and her influence. Agents wanted for the introduction and sale of this great work. Sent prepaid on receipt of price, $1.00. Address

[Pg 130]

Publications of the Health-Culture Co., 45 Ascension St., Passaic, N.J.

Health-Culture.

The largest and best illustrated monthly magazine published on the preservation and restoration of health, bodily development and physical culture for men, women and children. $1.00 a year; 10c. a number.

The Enlightened Life.

And How to Live It. By Dr. Latson; 365 pages, with portrait of the author. Cloth, $1.00.

This contains the leading editorials from Health-Culture, many of them revised and enlarged.

Common Disorders.

With rational Methods of Treatment. Including Diet, Exercise, Baths, Massotherapy, etc. By Latson. 340 pages, 200 illustrations. $1.00.

The Attainment of Efficiency.

Rational Methods of Developing Health and Personal Power. By Dr. Latson. Paper, 50c.; cloth, $1.00.

The Food Value of Meat.

Flesh Food Not Essential to Physical or Mental Vigor. By Dr. Latson. Illustrated. Paper, 25c.

Walking for Exercise and Recreation.

By Dr. Latson. 15c.

Dr. Latson's Health Chart.

Presenting in an Attractive and Comprehensive Form a Complete System of Physical Culture Exercises, fully Illustrated with Poses From Life, with Special Directions for Securing Symmetrical Development, for Building up the Thin Body, for Reducing Obesity, and for the Increase of General Vitality. 18×25 inches, printed on fine paper, bound with metal, with rings to hang on the wall. 50c.

Uncooked Food.

And How to Live on Them. With Recipes for Wholesome Preparation, Proper Combinations and Menus, with the Reason Uncooked Food Is Best for the Promotion of Health, Strength and Vitality. By Mr. and Mrs. Eugene Christian. Cloth, $1.00.

The New Internal Bath.

An Improved Method of Flushing the Colon or Administering an Enema. For the relief of Acute and Chronic Diseases. By Laura M. Wright, M. D. Illustrated. 25c.

[Pg 131]

Womanly Beauty.

Of Form and Feature. The Cultivation and Preservation of Personal Beauty Based upon Health and Hygiene. By Twenty Well-known Physicians and Specialists. With 80 half-tone and other Illustrations. Edited by Albert Turner. 300 pages, cloth and gold. Price, $1.00.

In this volume the Editor has brought together the teachings of those who have made a study of special features of the subject, and the result is a work that is unique and practical, not filled with a medley of receipts and formulas, so often found in books on beauty.

Manhood Wrecked and Rescued.

How Strength and Vigor Is Lost and How it may be Restored by Self-Treatment. A Series of Chapters to Men on Social Purity and Right Living. By Rev. W. J. Hunter, Ph. D., D. D. Cloth $1.00.

It contains the following chapters: The Wreck—An Ancient Wreck—A Modern Wreck—A Youthful Wreck—A Wreck Escaped—The Rescue Begun—The Rescue Continued—The Rescue Completed.

Illustrated Hints upon Health and Strength for Busy People.

Text and Illustrations by Adrian Peter Schimdt, Professor of Higher Physical Culture. Price $1.00.

The best System of Physical Culture published.

Courtship Under Contract.

The Science of Selection. A Tale of Woman's Emancipation. By J. H. L. Eager 440 pages, with portrait of the author. Price, $1.20 net. By mail, $1.30.

A novel with a purpose, higher than that of any other ever published, not excepting even "Uncle Tom's Cabin," as it aims to secure more of happiness in Marriage and the doing away with the divorce evil. The author presents, in the form of a clean, wholesome love story, some new ideas on the subject of Love, Courtship, Marriage and Eugenics.

Human Nature Explained.

A new Illustrated Treatise on Human Science for the People. By Prof. N. N. Riddell. Illustrated. 400 pages. Extra cloth binding, $1.00.

Men and women differ in character as they do in looks and temperament; no two are just alike. If you would know these "Signs of Character," read "Human Nature Explained," and you can read men as an open book. It gives the most complete system of reading character ever published.

Human Nature Indexed.

A Descriptive Chart for use of Phrenologists. By N. N. Riddle. 25c.

[Pg 132]

What Shall We Eat?

> The Food Question, from the Standpoint of Health, Strength and Economy. Containing Numerous Tables Showing the Constituent Elements of over Three Hundred Food Products and Their Relations, Cost and Nutritious Values, Time of Digestion, etc., Indicating Best Foods for all Classes and Conditions. By Alfred Andrews. Price, leatherette, 50c.; cloth binding. 75c.

The New Method.

> In Health and Disease. By W. E. Forest, B.S., M.D., Fellow of N. Y. Academy of Medicine. Sixteenth Edition. Revised and enlarged by Albert Turner, Publisher of Health-Culture. 350 pp., clo. binding, $1.

> It makes the way from weakness to strength so plain that only those who are past recovery (the very few) need to be sick, and the well who will follow its teachings cannot be sick, saving the need of calling a physician and all expenses for medicine.

Massotherapy.

Or the Use of Massage Rollers and Muscle Beaters in Indigestion, Constipation, Liver Trouble, Paralysis, Neuralgia and Other Functional Diseases. By W. E. Forest, M. D. 25c.

Constipation.

Its Causes and Proper Treatment Without the Use of Drugs. By W. E. Forest, M. D. The only rational method of cure. 10c.

Hygienic Cookery.

Or Health in the Household. By Susanna W. Dodds, M. D. $2.00.

It is unquestionably the best work ever written on the healthful preparation of food, and should be in the hands of every housekeeper who wishes to prepare food healthfully and palatably.

The Diet Question.

Giving Reasons Why—Rules of Diet. By Dr. Dodds. 25c.

The Liver and Kidneys.

With a Chapter on Malaria. Part I. The Liver and Its Functions, Diseases and Treatment. Part II. The Kidneys, Their Healthy Action and How to Secure It. Part III. Malarial Fever, Rational Treatment by Hygienic Methods. By Dr. Dodds. 25c.

Race Culture.

The Improvement of the Race through Mother and Child. By Susanna W. Dodds, M. D. Nearly 500 pages, $1.50.

Dr. Dodds' experience as a physician, teacher and lecturer has given her the preparation needed for the writing of this book. It is certainly safe to say that every woman, especially the mothers of young children and prospective mothers, should read it. No other work covers so completely the subject of health for women and children as in "Race Culture." [Pg 133]

Scientific Living.

For Prolonging the Term of Human Life. The New Domestic Science, Cooking to Simplify Living and Retaining the Life Elements in Food. By Laura Nettleton Brown. $1.00.

This work presents new views on the health question, especially as related to food. It treats of the life in food, showing that in the preparation of food by the usual methods the life-giving vitality is destroyed; that is, the organic elements become inorganic. The reason is clearly stated and recipes and directions for cooking, with menus for a balanced dietary, are given.

Cooking for Health.

Or Plain Cookery, With Health Hints. By Rachel Swain, M. D. $1.00.

This book is the outcome of progress in the kitchen, and provides for the preparation of food with direct reference to health. It is not an invalids' Cook Book, but for all who believe in eating for strength, and the use of the best foods at all times.

The No-Breakfast Plan and Fasting Cure.

By Edward Hooker Dewey, M. D. Cloth, $1.00.

Presents his theories in a clear, concise, practical way, together with specific and definite instructions for the carrying out of this method of living and treatment.

Experiences of the No-Breakfast Plan and Fasting Cure.

A letter in answer to the many questions asking for special details as to methods and result. By Dr. Dewey, 50c.

Chronic Alcoholism:

Its Radical Cure. A new method of treatment for those afflicted with the alcohol habit, without the use of drugs. By Dr. Dewey. 50c.

Health in the Home.

A Practical Work on the Promotion and Preservation of Health, with Illustrated Prescriptions of Swedish Gymnastic Exercises for Home and Club Practice. By E. Marguerite Lindley. $1.00.

Unquestionably the best and most important work ever published for the promotion of the health of women and children.

The Temperaments;

Or Varieties of Physical Constitution in Man in Their Relations to Mental Character and the Practical Affairs of Life, etc. By D. H. Jacques, M. D. Nearly 150 Illustrations. $1.50.

The only work published on this important and interesting subject. The author made it the special subject of study and was thoroughly familiar with all temperamental questions.

The Avoidable Causes of Disease;

Insanity and Deformity, Together with Marriage and Its Violations. By John Ellis, M. D. New Edition, Revised and Enlarged by the Author, with the Collaboration of Dr. Sarah M. Ellis. $1.00.

This book should be in every library, and if read and its teachings followed nearly all sickness and disease would be avoided with the accompanying suffering and expense—one of the most valuable works ever published.

Facial Diagnosis.

Indications of Disease as shown in the Face. By Dr. Louis Kuhne. Illustrated. $1.00. [Pg 134]

SCIENTIFIC LIVING

For Prolonging term of Human Life

The New Domestic Science, Cooking to Simplify Living and Retaining the Life Elements in Food.

By Laura Nettleton Brown.

A great truth is emphasized in this book, namely, that in the ordinary processes of cooking the organic elements become inorganic and food values are destroyed. This dietetic idea is most important, and it is claimed by the author that when generally known and made practical it will restore the racial vigor as nothing else can, free woman from the slavery of the cook stove and become a large factor in the solution of the servant problem.

The author does more than inform; she arouses and inspires; she also enters into the practical demonstration of the new way; food tables, recipes and menus are numerous and enlightening and will prove exceedingly helpful not only to busy housekeepers, but also to all persons who desire to get the greatest benefit and fullest enjoyment from the daily meals.

She refrains from urging the exclusive use of uncooked foods, but shows what kind of cooking can be made useful. A most interesting and practical feature of this work is the clear and discriminating instructions given for the application of heat in preparing food. From the author's point of view it becomes evident that the present mode of preparing food is not only unnecessarily laborious, but that it involves great waste of the raw material and puts a severe tax upon the digestive organs of the consumer.

The best thing about the new way to many minds, however, will be that it greatly enhances the appetizing qualities of the viands. It treats of the chemistry of food in a way that is easily understood and made practical. The concluding chapter of the book deals with "Associate Influences," and gives sound advice upon other factors than diet.

The volume is thoroughly sensible and enlightening; original without being cranky; radical without being faddish; withal, practical plain and entirely helpful. No one who is interested in the all-important question of scientific living can afford to be without this book. It will be found of interest to teachers and students of domestic economy. It is very carefully and thoroughly indexed, adding to its usefulness.

Printed on fine paper. Handsomely bound in extra cloth. $1.00 by mail on receipt of price. If not entirely satisfactory, money will be returned. Address

[Pg 135]

The New Internal Bath

The benefits and great importance of properly flushing the colon is now fully recognized and it has led to a large and increasing demand for syringes used for this purpose. The appliances in general use have one very serious fault, the water is discharged into the lower part of the rectum, which is distended, and thus produces an irritation which often proves injurious, causing and aggravating piles and other rectal troubles. It in frequently a cause of constipation and creates a necessity for continuing the use of enemas indefinitely.

Dr. Wright's New Colon Syringe

Consists of a strong, well made, four quart rubber bag or reservoir with two long Soft Rubber Flexible Tubes, by the use of which the water is easily carried past the rectum and into the sigmoid flexure, and by the use of the longest tube may be carried up to the transverse colon. The water is then discharged where it needed and the cleansing is made much more perfect than it can be in any other way. The tubing and the outlets are extra large, securing a rapid discharge of the water, which reduces the time required to less than one-half that usually taken, which is a very great advantage over other syringes. This new syringe will prove a most important help in the taking of "Internal Baths" in the "New Method" treatment as recommended by Dr. Forest and others, and will prove curative in many cases when all others fail.

Dr. Wright's manual on the taking of the "Internal Bath," containing full directions for its use in Constipation, Diarrhoea, Dyspepsia, Biliousness, Sick Headache, Kidney Troubles, Convulsions, Jaundice, Rheumatism, Colds, Influenza, La Grippe, Diseases of Women, Worms and Constipation in Children and other diseases, price 25c., is given free with each syringe.

Carefully packed in a fine polished wooden case, will be sent prepaid to any address on receipt of price, $5.00, with a copy of Dr. Forest's great work, "The New Method," the very best work on Health and Disease published. (Price, $1.00), both for $5.50.

An Infants' Flexible Rubber Tube will be sent for 75c. extra; New improved Vaginal Irrigator, $1.00; two Hard Rubber Rectal Tubes if desired, 25c extra. Agents wanted to introduce and sell this.

[Pg 136]

Health Culture Appliances

DR. WRIGHT'S COLON SYRINGE, for taking the New Internal Bath.

This consists of a one-gallon reservoir, one each, long and short flexible rubber colon tube, one box of antiseptic powder, and Dr. Wright's Manual of the New Internal Bath, all packed in a polished wooden case. Price, prepaid, $5.00.

THE PRIMO LADIES' SYRINGE. Price, $2.00. The only properly constructed Vaginal Syringe made.

Every woman should have a good syringe for use in emergencies and for purposes of cleanliness, which is essential to health, comfort and pleasure.

All women, married or single, should have a Primo. With each is sent full directions for use in all emergencies.

DR. FOREST'S MASSAGE ROLLERS.

These rollers are coming into general use wherever massage is needed and are a cure for many of the functional disorders as Dyspepsia, Constipation, Biliousness, Neuralgia, Rheumatism, Sleeplessness, Obesity, and wherever there is a lack of a good circulation of the blood; and the developers and facial rollers are used successfully for building up the form and the prevention of wrinkles and age in the face. The rollers consist of wheels about 1½ inches in diameter; around the centre is a band or buffer of elastic rubber.

No. 1, Body Roller, 6 Wheels, $2.—The best size for use over body, and especially for indigestion, constipation, rheumatism, etc.

No. 2, Body Roller, 4 Wheels, $1.50.—Smaller and lighter than No. 1, for small women it is best in size for use over the stomach and bowels, the limbs and for cold feet.

No. 3, Scalp Roller, $1.50.—Made in fine woods, and for use over the scalp, for the preservation of the hair.

No. 4, Bust Developer, $2.50.—The best developer made. By following the plain, physiological directions given, most satisfactory results can be obtained.

No. 5, Abdominal Roller, 12 Wheels, $4.—For the use of men to reduce the size of the abdomen and over the back.

No. 6, Facial Roller, $2.50.—Made in ebony; very fine for use over the face and neck, for preventing and removing wrinkles and restoring its contour and form.

No. 7, Facial Roller, $1.50.—Like No. 6. Made in white maple. In other respects the same.

No. 8, Abdominal Boiler, 8 Wheels, $3.50.—This is the same as No. 5, except with the less number of wheels. Is made for the use of women, for reducing hip and abdominal measure.

No. 1 Massage Vibrator, 24 Balls, price $2.00.

No. 2 Massage Vibrator, 12 Balls, price $1.25.

Dr. Forest's Manual of Massotherapy, containing nearly 100 pages, giving full directions for use, sent with each of the above.

TURKISH BATH CABINETS.

No. 1, a Double Walled Cabinet, the best made, with new and improved heater and manual giving full instructions for using the Cabinet for the Cure of Colds, Catarrh, Rheumatism, LaGrippe, Neuralgia, Kidney Trouble, Lumbago, Malaria, and many other disorders. Price $12.50.

No. 2 Cabinet Single Walled, with heater and instructions as above. Price $7.50.

DR. FOREST'S HEALTH CULTURE VASELINE SPRAY and Bottle of Catarrh Remedy. Price, $2.00.

THE WILHIDE EXHALER. Price $1.00.

Special descriptive circulars of any of the above sent on application.

Address all orders to

[Pg 137]

Uncooked Foods And How to Use Them.

With recipes for wholesome preparation, proper combinations and menus, with the reason why it is better for the promotion of health, strength and vitality to use uncooked than cooked foods, by Mr. and Mrs. Eugene Christian, with an Introduction by W. R. C. Latson, M. D.

It will meet a widespread want filled by no other work that has ever been published, and will do very much to solve the question of how to live for health, strength, and happiness.

It will simplify methods of living—help to solve the servant question and financial problems, as well as point the way for many to perfect health. The following chapter headings show something of the scope and value of this.

CONTENTS.

PART FIRST—

- Why This Book Was Written,
- Introduction,
- The Emancipation of Women,
- The Functions of Foods,
- Food Products,
- Selection of Foods,
- Raw Foods,
- Preparation of Foods,
- Preparation of Uncooked Wood,
- Effects of Cooking Food,
- Tables Giving Nutritive Values, etc.
- Food Combinations,
- Condiments,
- Bread—Fermentation,
- Economy and Simplicity,
- As a Remedy.

PART SECOND—

- How to Begin the Use of Uncooked Foods.
- Recipes for—
- Soups,
- Salads (35 kinds),
- Eggs, Meat and Vegetables,
- Cereals,
- Bread, Crackers and Cakes,
- Nuts,
- Fruits and Fruit Dishes,
- Evaporated Fruits,
- Desserts,
- Jellies and Ices,
- Drinks,
- Menus,
- Miscellaneous.

It is the most important work on the food question ever published. Bound in cloth. Price, $1.00; with a year's subscription to Health-Culture, $1.50. Address,

[Pg 138]

COMMON DISORDERS

Including Diet, Exercise, Baths, Exercise, Massotherapy, Etc.

BY W. R. C. LATSON. M. D.

This is a practical handbook and guide for the home treatment of the sick without the use of drugs, with suggestions for the avoidance of disease and the retaining of health and strength. A book for those who would get well and keep well.

CONTENTS.

Introduction.—What the Body Is. Cell Life and Its Construction. Circulation of the Blood and What It Is. What Exercise Does.

Massage. Principles and Practice. How It Acts as a Remedy.

Massotherapy. Showing How It Is Applied.

Special Exercises. Including Those for Development and Remedial Work.

Tissue Building. Special Diet, with Menus.

Obesity. Its Cause and Treatment Instructions for General Reduction.

Indigestion. Causes of Dyspepsia. What to Do to Secure Good Digestion.

Constipation. Its Causes. Treatment by Hygienic Measures.

Rheumatism. Muscular and Articular. Treatment.

Gout. Causes. Symptoms. General and Local Treatment.

Neuralgia. Causes and Symptoms. The Only Rational Treatment.

Sprains and Synovitis. Symptoms. Treatment.

Varicose Veins and Swollen Glands. The Cause and Treatment.

Baldness. Treatment for Restoring the Hair.

Lung Disorders. How to Improve Breathing. The Prevention and Treatment of Consumption.

Round Shoulders and Protruding Collar Bones. How to Overcome Them, with Special Exercises.

How to Strengthen the Back. The Cause of Spinal Weakness.

How to Strengthen the Trunk. The Importance of Strong Bodily Muscles.

A Chair as a Gymnasium. How to Use a Bedroom Chair as a Complete Gymnasium Apparatus.

The Hygiene of the Skin. Nerves of the Skin. Sun Baths.

Modern Nervousness. The Best Treatment.

Smallpox. Its Nature. Prevention. Treatment of Smallpox.

Sunstroke. Causation and Treatment. How to Avoid It. What to Do When Prostrated.

In this work the author sets forth the methods he has pursued and found be practical and successful. Over 300 pages and 200 Illustrations. Price $1.00.

[Pg 139]

RACE CULTURE

THE IMPROVEMENT OF THE RACE THROUGH MOTHER AND CHILD. By Susanna W. Dodds, M. D.

A large 12mo. volume bound in extra cloth, price, $1.50

The time has come when parents must consider the responsibilities that rest upon them in relation to their children and make a study of Eugenics. This cannot be avoided or shirked and especially should prospective mothers study the subject in all its bearing, and know what you should do and what you should not do to insure the best possible for your unborn child. What conditions will promote the best for health, and afford the highest degree of intellectual and moral development. What limit you shall place upon the number of children. Race Suicide is not so serious a question as Race Culture, which may be easily attained by giving proper attention to the subject.

The author of "RACE CULTURE" has made a most careful study of the whole subject, starting from the foundation, taking up pre-natal culture in all its bearings, including the marriage relations and the father's responsibilities. Considering the health and the well-being of the prospective mother and her diseases. How childbearing may be made easy, the first care of and the feeding of the babe, all the diseases of infancy and childhood and their treatment without the use of drugs.

The avoidable causes of disease in children and adults are fully considered and a voluminous appendix treats of the use of water, massage, exercise, food and drinks, and how to prepare them as remedial agencies.

It is safe to say that no greater or more important work on this subject has ever been written.

Every woman and especially every prospective mother should read it. Its cost is as nothing compared to its value. Price, $1.50 by mail. Address

[Pg 140]

The Food Value of Meat

Flesh Food Not Essential to Mental or Physical Vigor.

By W. R. C. LATSON, M. D.,

The most valuable work on Practical Dietetics that has been published. The Food Question is considered in its relation to health, strength and long life. Some idea of the scope may be seen from the following

CONTENTS

INTRODUCTION. Importance of the Subject. Influence of Foods on the Health and Morality of the Community. The Most Important Question of Dietetics. Classes of Foods. Description of Proteids. The Starches. Conversion of Starches into Sugars. Fruit Sugar. The Fats. Salts. Effect of Cooking Upon Foods.

DIGESTION. Definition of the Process. Saliva. The Ptyalin. Effect of Eating Sugar with Starchy Foods. Gastric Digestion. The Stomach; The Gastric Juice; Peptones; Digestion In the Intestines; Importance of Digestion; Tabular Statement of the Digestive Process.

COMPOSITION OF FOODS. The Four Elements of Food; Proper Proportion of Each Element; Selection of Balanced Foods; Table of Food Analyses; Value of Cooked Vegetables; The Reason Why Many Vegetarians Fail; Fresh Fruits; Pure Water; The Grains; The Legumes; Nuts.

FOOD VALUES OF FLESH MEATS. The Question at Issue; Biological Data, What They Indicate; The Intestinal Tract; The Food Value of Meat; Poisons; Disease Infection; The Strongest Argument Against the Use of Flesh Meat; Vigorous Vegetarians; Intellectual Vegetarians; Vegetarianism and Vigor.

COMBINATIONS OF FOODS. Principles; Cooked and Uncooked Foods; Model Menus; Breakfast; Luncheon; Dinner; Advantages of Vegetable Foods.

Price by Mail, in Paper. 25c, Cloth Binding, 50c.

COMMON DISORDERS

Causes, Symptoms, and Hygienic Treatment, by the use of Water, Massotherapy, and other Rational Methods.

By W. R. LATSON, M. D.

Among the diseases considered may be mentioned Indigestion, Constipation, Rheumatism, Neuralgia, Lung Troubles, Gout, Nervousness and other minor complaints. The work contains nearly 300 pages, profusely illustrated. Bound in Cloth. Price, $1.00. Sent by mail on receipt of price.

[Pg 141]

The Up-to-date Woman

needs to know something more than simply How to Cook and follow recipes brought to her attention in Cook Books

SHE SHOULD KNOW

- What are the Best Foods for her family.
- What Foods will keep all Well and Strong.
- What is best for the Children.
- What do the Men Need.
- What Foods are Economical and Nutritious.
- What are best Food Combinations.
- How often is Meat Necessary.
- What are the Best Meat Substitutes.
- What is the Food Value of Fish.
- What is the Food Value of Milk.
- What is the Food Value of Nuts.
- Are Beans Nutritious and Healthful.
- Is Nut Butter better than Cow Butter.
- Are Tea and Coffee Injurious.
- Which Food Digests Quickly and which Slowly.

- How to Get the Most Food Value for the Least Money.

    All these and many other questions are answered in
    Prof. Andrews Great Book

What Shall We Eat?

The Food Question from the standpoint of Health, Strength and Economy. Indicating Best Foods for all Classes and Conditions.

This work covers every phase of the food question in a practical way.

Shows how food is digested and gives the constituent elements of all food products, their cost, food values, time of digestion, etc., Comparative value of beef, mutton, pork, eggs, fish, fowl, oysters, the grains, breads, peas, beans, milk, butter, cheese, sugar, beer, fruits, nuts, etc., which make flesh, bone, nerve; which gives most for least money. 25 tables showing results of nearly 1500 food analyses. Price in leatherette binding, 50 cents, cloth 75 cents, postpaid.

If not satisfied money promptly returned. Every man should order this for his wife, or some other woman. Send stamps.

[Pg 142]

The Enlightened Life and How to Live it

By W. R. C. LATSON, M. D.

Author of "Common Disorders," "The Attainment of Efficiency," "Food Value of Meat," Etc.

This work contains a collection of Dr. Latson's strong editorials that have appeared in Health-Culture, carefully revised and enlarged, with other matter. The great interest that has been manifested in these leaders will insure a demand for this work. The scope will be seen from the following chapter headings:

Introduction—The Ultimate Ideal—The Mind and Its Body—What Shall a Man Take in Exchange for His Soul?—Health as an Asset—The Waste of Life—Health as a Factor in Business Success—The Causation of Disease—Are Weakness and Disease Increasing?—The Detection of Disease—The Prevention of Disease—Heredity and Disease—Disease: Its Nature and Conquest—Methods of Healing—Drug Medication in the Treatment of Disease—Religion and Medicine—Worry the Epidemic of the Day—Race Suicide—"Race Suicide," Pro and Con—Simplified Living—The Death-Dealing Detail—The Slaughter of the Innocents—Crimes Against Children—Sleep and Rest—Mental and Physical Effects of Music—The Common Sense of Foods and Feeding—The Mission of Pain—Drugs—The Surgical Operation Frenzy—Vaccination; Blessing or Curse?—Free Water Drinking as a Hygienic Measure—Evil Effects of Alcohol—The Pinnacles of Absurdity.

Published in large, clear type, handsomely bound in cloth. Price, sent prepaid, $1.00. Address

[Pg 143]

The Health Culture Magazine

ELMER LEE., A. M., M. D., EDITOR

PRINCIPLES AND OBJECTS

Health Culture seeks the advancement of humanity by declaring the obvious teachings of nature.

Health Culture aims to educate the people out of superstition, misunderstanding and fear arising from the imperfect interpretation of natural principles.

Health Culture recognizes that health and comfort, happiness and long life are desirable and attainable by the faithful observance of hygiene. That neglect and abuse of natural and simple living inevitably leads to weakness, degeneracy, disease and death.

Health Culture from the scientific sense as well as on grounds of sentiment opposes the taking of life needless to obtain food for man.

Health Culture holds that food products of the vegetable kingdom are ample and favorable for a safe, complete and full development of the kingdom of man.

Health Culture opposes as needless and wasteful of life those research activities known as vivisection, also as contrary to human interest the use of drugs, serums, vaccines and chemicals as medicines or preventives of disease by legal compulsion.

Health Culture is an illustrated Monthly, Standard Magazine size; $1.00 a year, 15 cents a No., Canadian subscriptions $1.25, Foreign $1.50.

Address The Health Culture Co., Passaic, N. J.

www.ingramcontent.com/pod-product-compliance
Lightning Source LLC
Chambersburg PA
CBHW031426210526
45464CB00005B/2067